# AS/A2 UNIT 3

## STUDENT GUIDE

## CCEA

# Biology

## Practical skills in biology

John Campton

HODDER
EDUCATION
AN HACHETTE UK COMPANY

Every effort has been made to trace all copyright holders, but if any have been inadvertently overlooked, the Publishers will be pleased to make the necessary arrangements at the first opportunity.

Orders: please contact Hachette UK Distribution, Hely Hutchinson Centre, Milton Road, Didcot, Oxfordshire, OX11 7HH. Telephone: +44 (0)1235 827827. Email education@hachette.co.uk. Lines are open from 9 a.m. to 5 p.m., Monday to Friday. You can also order through our website: www.hoddereducation.co.uk

© John Campton 2018

ISBN 978 1 5104 1915 5

First published in 2018 by
Hodder Education,
An Hachette UK Company
Carmelite House
50 Victoria Embankment
London EC4Y 0DZ
www.hoddereducation.co.uk

Impression number 6

Year 2022

This guide has been written specifically to support students preparing for the CCEA AS and A-level Biology examinations. The content has been neither approved nor endorsed by CCEA and remains the sole responsibility of the author.

Cover photo: Alexander Raths/Fotolia; other photos: **p. 17** Biophoto Associates/SPL, **p. 20** Power and Syred/SPL, **p. 25** Steve Gschmeissner/SPL, **p. 27** and **p. 69** Dr Keith Wheeler/SPL, **p. 57** and **p. 58** Dr Jeremy Burgess/SPL, **p. 66** and **p. 101** M. I. Walker/SPL

Typeset by Integra Software Services Pvt. Ltd, Pondicherry, India

Printed and bound by CPI Group (UK) Ltd, Croydon, CR0 4YY

Hachette UK's policy is to use papers that are natural, renewable and recyclable products and made from wood grown in well-managed forests and other controlled sources. The logging and manufacturing processes are expected to conform to the environmental regulations of the country of origin.

# Contents

# AS practical skills

## Skills Guidance

## Questions & Answers

# A2 practical skills

## Skills Guidance

## Questions & Answers

# ■ About this book

This book will help you develop the skills in practical biology required as part of CCEA's A-level Biology specification. Since CCEA's A-levels consist of AS and A2, the book essentially contains two guides: the first is for AS Biology, followed by another for A2 Biology. The guide for each level contains details of the practicals that you should know, the mathematical skills associated with processing the data generated in practical work and an explanation of how practical skills will be assessed with exemplar questions and answers.

Each guide includes all the types of **practical work** that you are expected to experience (these are italicised in the specification for Units 1 and 2). For each practical, there are *practical tips*, *exam tips* and *knowledge checks* in the margin. Answers to the knowledge checks are provided towards the end of the book. You will also find *definitions* for some key terms in the margin.

Since practical work, both in the laboratory and in the field, often generates quantitative results, each guide also describes the **mathematical skills** required in dealing with data. Understanding of the mathematical skills specified at AS is also required by students at A2; while understanding of *logarithms and statistics* is confined to A2.

While some questions in the Unit 1 and Unit 2 written exams may assess practical skills, they are mostly assessed in Unit 3, at both AS and A2. Unit 3 consists of two components. Since implementation skills cannot be assessed in a written exam, a series of **practical tasks** will be assessed by your teachers — for each task undertaken, evidence must be provided for the level achieved. There is also a **practical skills written paper** which assesses your understanding of practical skills and your ability to apply them to familiar and unfamiliar contexts.

The **Questions & Answers** section presents questions that cover most areas of the specification. There are answers written by two students of differing ability with comments on their performances and how they might have been improved. These will be particularly useful during your final revision. There is a range of question styles that you will encounter in the Unit 3 exam, and the students' answers and comments should help with your examination technique.

Ensure you access the biology specification at www.ccea.org.uk.

## Hazard and risk

All practical tasks described in this guide should be risk assessed by a qualified teacher before being performed either as a demonstration or as a class practical. Safety goggles and a laboratory coat must be worn where it is appropriate to do so. The author and the publisher cannot accept responsibility for safety.

# AS practical skills

## ■ AS Unit 1 practical work

### Biological molecules
### Tests for carbohydrates and proteins

You should be familiar with the tests summarised in Table 1.

**Knowledge check 1**

Which biochemical test requires heat to give a positive result?

**Table 1** Biochemical tests for carbohydrates and proteins

| Test | Biochemical tested | Relative proportions of reagent(s) and test solution | Colour change if biochemical present — positive result |
|------|-------------------|------------------------------------------------------|--------------------------------------------------------|
| Iodine test | Starch | Few drops added to test solution; or sample of test solution added to dilute iodine | Straw yellow or orange (depends on concentration) to **blue-black** |
| Benedict's test | Reducing sugars — all monosaccharides and some disaccharides (e.g. maltose) | Approximately equal volumes of Benedict's solution and test solution. These are then heated in a water bath | Blue to **green, yellow, orange and finally brick-red precipitate** (colour and amount of precipitate dependent on quantity of reducing sugar present) |
| Glucose-specific tests, e.g. Clinistix or Diastix | Glucose | Test-strip is dipped in test solution (for 10 seconds) | Colour changes depend on the test-strip used — amount of glucose present estimated by comparing with standards shown on the packaging |
| Biuret test | Proteins (soluble or globular) | 1 Potassium hydroxide is added to test solution until it clears, and then a drop of copper sulfate added down side of tube 2 Biuret reagent (contains mixture of above) added to test solution | 1 **Blue ring** at surface and then, on shaking, the solution turns **purple** 2 Blue to **violet** (or **mauve** or any shade of **purple**) |

The disaccharide sucrose is not a reducing sugar. Its presence can be detected if it is first hydrolysed (into glucose and fructose) by heating (in a water bath) with dilute hydrochloric acid (HCl); the acid is then neutralised with sodium hydrogencarbonate ($NaHCO_3$) before being tested with Benedict's solution. To ensure the validity of concluding its presence, two further tests must be carried out: the original solution must undergo a Benedict's test to ensure that reducing sugar was not also present; and, since this procedure would also work with starch (hydrolysed into glucose), its absence must first be determined by testing with iodine solution.

**Validity** How well a test measures what it is supposed to measure. You cannot make valid conclusions if there is serious doubt about the precision, accuracy and/or reliability of the results.

# Analysis of amino acids using paper chromatography

Paper chromatography is a technique for separating and identifying substances extracted (in a suitable solvent) from biological material. The technique for amino acids involves:

1   **Preparation of chromatography tank.** Pour solvent into the tank to a depth of about 3 cm and put the lid on so that the atmosphere becomes saturated with vapour.

2   **Preparation of chromatogram.** Having washed your hands to remove amino acids in sweat and handling the chromatography paper only at the sides, or alternatively wearing disposable gloves, draw a pencil line across it about 5 cm from one end; using a micro-pipette, place a small drop of the amino acid mixture at a marked point on this origin line, let it dry, place another drop on top of the first and dry again, and repeat this process to concentrate the amino acids while keeping the spot as small as possible.

3   **Running the chromatogram.** Suspend the prepared chromatography paper in the tank, ensuring that the pencil origin line is above the surface of the solvent (see Figure 1); allow the solvent to run up the paper for some hours during which time the amino acids will be carried different distances (according to their solubility in the solvent) and so be separated; before the solvent arrives at the opposite end, the chromatography paper is removed and the solvent front drawn in pencil; the chromatogram is dried.

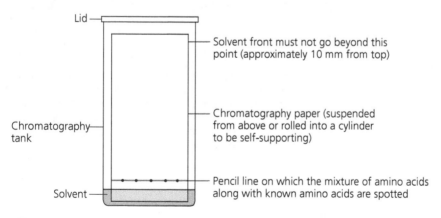

Lid

Solvent front must not go beyond this point (approximately 10 mm from top)

Chromatography paper (suspended from above or rolled into a cylinder to be self-supporting)

Chromatography tank

Pencil line on which the mixture of amino acids along with known amino acids are spotted

Solvent

**Figure 1** Paper chromatography apparatus for the analysis of amino acids

4   **Developing the chromatogram.** In a fume cupboard, and wearing gloves, the dry chromatogram is sprayed with dilute ninhydrin reagent; this is again dried and heated (say, in an oven) so that purple spots indicating the location of amino acids appear; the spots are encircled with a pencil and a horizontal line drawn through the middle of each spot.

**5** **Identifying the amino acids.** For each amino acid, the distance from the origin line to the middle of the spot and to the solvent front is measured, and the $R_f$ value (relative front value) calculated as:

$$R_f = \frac{\text{distance moved by spot}}{\text{distance moved by solvent front from origin line}}$$

The amino acids may be identified from a table of $R_f$ values, though it is preferable to compare the spots in the mixture with known amino acids run in the same chromatogram. This is because $R_f$ values are specific for the type of paper used, the solvent and the temperature. The colour of the spots, which may vary from brown to purple, may also be used in identification (and uniquely proline produces a yellow spot).

However, just because two spots travelled the same distance and have the same colour is not a guarantee that they are the same amino acid. Different amino acids can have very similar $R_f$ values for any particular solvent. This problem is solved by using two-way chromatography: the amino acids are further separated by turning the paper through 90° and repeating the process at right angles to the original using a different solvent.

There are a large number of precautions in using paper chromatography; for example, pencil must be used to mark the chromatogram (since it is inert), while to obtain a concentrated yet small spot, the spot must be dried before the spotting process is repeated.

# Enzymes
## Properties of enzymes

Enzyme activity is affected by a number of factors:
- temperature
- pH
- substrate concentration
- enzyme concentration

## Use of water baths to change or control temperature

There are two types of heated water bath that you may use.
- You can use a beaker of water heated, using a Bunsen burner, to the desired temperature, checking this with a thermometer. Since the water will cool down, it must be heated intermittently to just above the temperature needed and then allowed to fall to just below this level. With regular attention, the temperature can be maintained to within ±2°C.
- You could use a thermostatically controlled water bath. It does not keep the temperature completely constant but is more precise than heating a beaker of water. The temperature should be checked with a thermometer, initially for establishing the desired temperature and then regularly thereafter to monitor any fluctuations.

A low temperature, such as 10°C, can be achieved by using a mixture of water and ice in a beaker. The temperature is monitored using a thermometer and adjusted by adding more water or ice as appropriate.

**Exam tip**

Some biologists measure to the *front edge* of the spot on the chromatogram (rather than the middle). You must do whichever is indicated on the exam paper.

**Knowledge check 2**

An amino acid is spotted on a line that is 30 mm from the end of the paper. After running and developing the chromatogram, the amino acid spot is found to be 84 mm from the end and the solvent front 150 mm from the end. Calculate the $R_f$ value for this amino acid.

Precaution An action taken in advance to ensure safe procedures and reduce any sources of error.

**Knowledge check 3**

Describe a precaution in developing the chromatogram.

## Use of buffers to change or control pH

The pH scale is a measure of the acidity or alkalinity of a solution. It is an indication of the concentration of hydrogen ions relative to water. The pH of water is 7. Solutions with a higher concentration of hydrogen ions than water have a lower pH and are acidic. Solutions with a lower hydrogen ion concentration than water have a higher pH and are alkaline. The pH scale is shown in Figure 2.

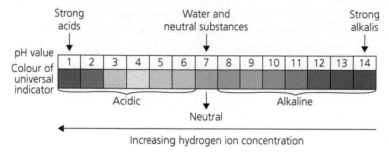

**Figure 2** The pH scale

Adding just a tiny amount of a strong acid (or strong alkali) to a neutral solution changes its pH dramatically. Buffers are solutions that resist this change and help to maintain a constant pH. Buffers can be prepared to maintain the pH of a solution at any given value.

## Use of different concentrations of solutions

To prepare different concentrations of solutions it is usual to prepare a stock solution, say a 10% solution (by dissolving 10 g of solid in 100 cm$^3$ of water). There are two types of dilution.

- One type of dilution produces an arithmetic series of concentrations: for example, 10%, 9%, 8%, 7% etc. Table 2 shows how you would prepare an arithmetic series.

**Table 2** Preparing an arithmetic series of dilutions

| Volume of 10% stock solution/cm$^3$ | 10 | 9 | 8 | 7 | 6 | 5 | 4 | 3 | 2 | 1 |
|---|---|---|---|---|---|---|---|---|---|---|
| Volume of water/cm$^3$ | 0 | 1 | 2 | 3 | 4 | 5 | 6 | 7 | 8 | 9 |
| Concentration of solution/% | 10 | 9 | 8 | 7 | 6 | 5 | 4 | 3 | 2 | 1 |

- Another type of dilution produces a logarithmic series of concentrations. This is serial dilution: each solution along the series is less concentrated than the previous one by a set factor. Figure 3 shows how you would undertake a doubling dilution so that subsequent dilutions are half the concentration of the previous solution.

Other dilution factors can be used, commonly ×10. To get a ten-fold dilution, you add 1 cm$^3$ stock solution to 9 cm$^3$ water, 1 cm$^3$ of this dilution to 9 cm$^3$ water, and so on. Serial dilution produces a wide range of dilutions — the doubling dilution, shown in Figure 3, produces a range of 10% to 0.156% solutions — and the bigger the dilution factor the greater the range. You need to be careful when plotting this range on a graph.

Sometimes dilutions are made with buffer solutions so that the pH of each dilution is the same.

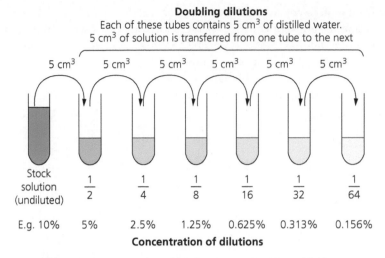

**Doubling dilutions**
Each of these tubes contains 5 cm³ of distilled water.
5 cm³ of solution is transferred from one tube to the next

**Figure 3** Preparing a serial dilution — doubling dilutions

There are many enzyme systems that may be used to investigate the influence of these factors. Investigations of the listed factors are shown below, each with a different enzyme system.

## Effect of temperature on the activity of lipase

Phenolphthalein is an indicator that is pink in alkaline solutions of about pH 10. When the pH drops below pH 8.3, phenolphthalein goes colourless. In this experiment an alkaline solution of milk, lipase and phenolphthalein will change from pink to colourless as the fat in the milk is hydrolysed to produce fatty acids thus reducing the pH below 8.3.

Milk is mixed with sodium carbonate solution to make the solution alkaline and 5 drops of phenolphthalein are added.

A series of water baths are set to a range of temperatures. Each water bath has placed in it a test tube containing the milk mixture and a small beaker of lipase. When the temperature of the solutions has equilibrated to the temperature of the water bath (say, after 10 minutes) lipase is added to the test tube and a stopwatch started (see Figure 4).

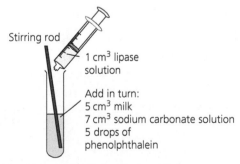

Stirring rod

1 cm³ lipase
solution

Add in turn:
5 cm³ milk
7 cm³ sodium carbonate solution
5 drops of
phenolphthalein

Stir and start timing when lipase added

**Figure 4** Apparatus for investigating the action of lipase on fats in milk

The time taken for the mixture to turn from pink to colourless is a measure of enzyme activity. The volumes and concentrations are kept constant — these are the controlled variables. Temperature is the independent variable (IV). The dependent variable (DV) is 'time taken', which may be converted to a rate of reaction by calculating its reciprocal. In any case, a short time indicates a high rate of reaction, which you would expect to be optimal at approximately 40°C. Results are recorded in a table and plotted on a graph (see Figure 5).

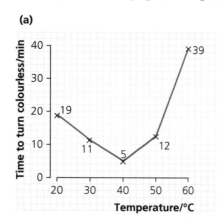 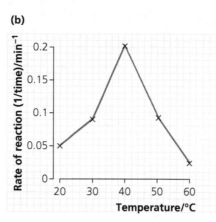

**Figure 5** The effect of temperature on the action of lipase, in terms of (a) time to turn colourless and (b) rate of reaction

*Interpretation* of the results will require you to explain an increase in lipase activity up to the optimum temperature with reference to kinetic theory (molecules moving more rapidly and so colliding more frequently) and a decrease above the optimum as a result of lipase denaturation (specific bonds holding the tertiary structure having been broken).

In *evaluation*, you would refer to the subjective nature of determining 'when the solution becomes colourless'. Reliability of the results will be increased by replication. Replication may be provided by including the results of other students — this, of course, compounds the subjectivity of measuring when the solutions have become colourless. It would be possible to vary the concentration of the lipase to investigate the effect of enzyme concentration. However, this experiment cannot be used to assess the effect of different pHs, since the mixture has to be alkaline initially.

## Effect of pH on the activity of amylase

Amylase solutions are mixed with different pH buffers and then solutions of starch added to provide a series of reaction mixtures. Measure the time taken for amylase to completely break down starch, by withdrawing samples of each reaction mixture at 10-second intervals and noting the time at which the solution no longer gives a blue-black colour with iodine solution (which should remain orange). The rate of reaction can be calculated as the reciprocal of 'time' (i.e. 1/time).

### Exam tip

When calculating the reciprocal, use 100/time or even 1000/time so that you do not get a rate with too many decimal places.

**Controlled variables** Variables that might influence the results and so need to be kept constant, so that the results are only influenced by the independent variable.

**Independent variable** The variable that is changed or manipulated in an experiment to test the effects on the dependent variable.

**Dependent variable** The variable being tested and measured in an experiment.

### Knowledge check 4

Explain why the solution of lipase is placed in the water bath for 10 minutes before adding $1\,cm^3$ to the milk (forming a reaction mixture).

**Reliability** The degree to which an experiment produces stable and consistent results.

### Knowledge check 5

Identify the independent and dependent variable in this experiment.

*Interpretation* should refer to an optimum rate of activity at, or near, a particular pH. As the pH differs from this optimum, changes in the hydrogen ion concentration alter charges on the amino acids that make up the active site of the enzyme. As a result, substrate molecules will less readily bind. At more extreme pHs, bonds that maintain the tertiary structure are broken so the enzyme is denatured.

In *evaluating* the experiment, the main errors will relate to timing: that the stopwatch is started immediately the starch is mixed with the enzyme–buffer solution; and that there is no delay in sampling. A delay in sampling would cause the reaction time to be underestimated (and the rate to be overestimated). There are other issues worth discussing: there may be some difficulty in determining when a blue-black colour is no longer formed; there will be a lack of precision in determining reaction time since samples are only taken every 10 seconds, rather than more frequently.

> **Knowledge check 6**
>
> State which variables you would need to control in this experiment.

> **Knowledge check 7**
>
> Why is it important to add amylase to the buffer prior to adding the starch solution?

## Effect of substrate concentration on the activity of catalase

The breakdown of hydrogen peroxide to water and oxygen is catalysed by the enzyme catalase. This enzyme is present in all living tissues, as hydrogen peroxide is a toxic by-product of metabolism. Catalase activity can be investigated in the laboratory by determining the amount of oxygen produced in a period of time (say, 30 seconds). A series of different concentrations of hydrogen peroxide are prepared and added to a source of catalase, such as puréed potato.

There are different methods available for measuring oxygen production over this time period and each has its advantages:

- Catalase solution is added, via a syringe, to hydrogen peroxide in a measuring cylinder and the height of the foam generated measured — Figure 6(a).
- Catalase solution is placed in a flask, hydrogen peroxide added via a burette and the oxygen produced collected by displacing water — Figure 6(b).
- Catalase solution is placed in a flask, hydrogen peroxide added via a burette and the oxygen produced collected in a gas syringe — Figure 6(c).

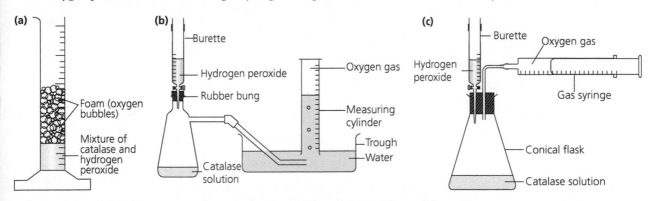

**Figure 6** Measuring the amount of oxygen produced (a) as height of foam, (b) oxygen collected by displacing water and (c) oxygen collected in a gas syringe

The first method, measuring the height of the foam, is easier to set up (and can readily be repeated) but it may not produce sufficiently accurate results: some of the oxygen produced will escape; and there may be difficulty in measuring the level to which the foam rises. In the other two methods none of the oxygen should escape. Method two, collecting gas by displacing water, involves apparatus which is more difficult to set up and so taking repeat measurements will be time-consuming.

*Interpretation* of results requires you to explain that as substrate concentration increases then collision with an enzyme is more likely and so the rate of reaction increases. However, at high substrate concentrations there are insufficient enzyme molecules for substrate molecules to bind to, the enzymes become saturated and so the rate of reaction reaches a maximum. In *evaluation* most concern will relate to the measurement of the oxygen produced: oxygen may escape, especially in the 'measurement of the height of foam' method, and possibly from the connections in the other methods if these are not tight; a little oxygen may also dissolve in the water in the 'displacement of water' method.

## Effect of enzyme concentration on the activity of protease

Gastric protease (also known as pepsin) initiates the digestion of proteins to peptides in the stomach. To investigate the effect of enzyme concentration a series of different protease concentrations is prepared: 1%, 2.5%, 5%, 7.5%, 10%. These are further diluted when $1\,cm^3$ of each protease solution is added to $9\,cm^3$ of dilute hydrochloric acid (of concentration $0.01\,mol\,dm^{-3}$). The dilute acid provides a pH of 2, the optimum pH for gastric protease, while the enzyme concentration now ranges from 0.1% to 1%.

The substrate used is albumen, the protein in egg white. Egg white is introduced into each of five small tubes (of known length) which are then heated in boiling water to solidify the albumen (denaturing it and making it more readily digested). One small tube of boiled egg white is added to each test tube containing a different protease concentration, and left for 24 hours (see Figure 7).

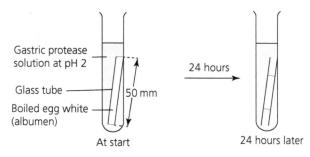

**Figure 7** Apparatus for investigating the effect of gastric protease on albumen

The next day, the length of the egg white column is measured and the length digested determined. The results are tabulated and plotted on a graph as 'length of egg white digested' against concentration of protease solution.

**Knowledge check 8**

The breakdown of hydrogen peroxide is an exothermic reaction generating heat. How would this affect the results?

**Knowledge check 9**

How would you keep temperature constant in this experiment?

**Knowledge check 10**

Describe and explain the results that would be expected in this experiment.

**Knowledge check 11**

Gastric protease can be used to investigate the effect of pH on enzyme activity. In such an experiment how would you vary the pH?

# Demonstration of enzyme immobilisation

The enzyme sucrase may be entrapped in calcium alginate beads by following the gel immobilisation procedure: sodium alginate is mixed with a sucrose solution and, using a syringe, is dropped into a solution of calcium chloride to form beads. The use of immobilised sucrase is shown in Figure 8.

Sucrose has no effect on a glucose-specific test strip but after passing through the column of alginate-entrapped sucrase, the test proves positive. This is because one of the products of sucrose hydrolysis is glucose.

This apparatus may be used to investigate how factors (flow rate, number of beads in the column and bead size) affect the rate of reaction.

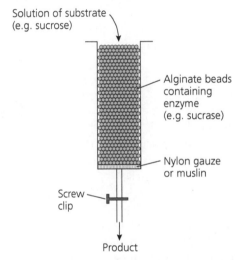

**Figure 8** Apparatus used for demonstrating enzyme immobilisation

# Using a colorimeter to follow the course of a starch–amylase reaction

To follow the course of an enzyme-controlled reaction, known amounts of substrate and enzyme are mixed. Then, at regular intervals, the amount of substrate which has not yet been acted on is measured.

## Procedure

$10\,cm^3$ 1.5% starch solution and $5\,cm^3$ 0.5% amylase solution are allowed time to equilibrate to a constant temperature in a water bath, and then mixed. Immediately (to obtain a result at time 0) a $1\,cm^3$ sample of the mixture is withdrawn using a syringe and added to $10\,cm^3$ dilute iodine solution. At 1-minute (or 30-second) intervals, further $1\,cm^3$ samples are removed and added to dilute iodine solution.

## Using a colorimeter

A colorimeter is set up to measure starch content of samples taken from a reaction mixture of starch and amylase. Any starch left will give a blue-black colour: a dark colour means that a lot of starch remains, a light colour means there is relatively little. The intensity of blue-blackness is measured using a colorimeter (see Figure 9).

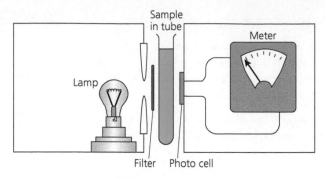

**Figure 9** A colorimeter

**Practical tip**

The colorimeter measures both *transmission* (amount of light passing through the solution) and *absorbance* (amount of light absorbed). Transmission is more commonly used, as absorbance involves a logarithmic scale.

To ensure that readings accurately reflect the colour density of the solution, the cuvettes used must be clean and free of any surface marks. It is preferable to use the same cuvette, rinsing and drying it between samples, and ensure that it is orientated in the same way when inserted into the colorimeter.

## Using a filter

A solution is blue because it transmits blue light and absorbs other colours, most efficiently red. So to measure the intensity of blueness a red filter is used in the colorimeter. Generally then, a filter is chosen which is at the opposite end of the spectrum to the colour of the solution being measured. This will maximise any change in colour and provide the greatest range in colorimeter readings.

**Knowledge check 15**

Explain why a red filter is used in a colorimeter when measuring the 'blueness' of a solution.

## Standardising the colorimeter

Immediately before measuring the blueness of each sample, a colorimeter tube (or cuvette) containing only dilute iodine solution (sometimes called the blank) is placed in the colorimeter and the instrument adjusted to 100% transmission. This standardises the instrument so that subsequent readings are comparable. With the use of the colorimeter you then have a series of readings for the samples of the reaction mixture taken at different time intervals.

## Constructing a calibration curve

These values can be plotted as % transmission against time. However, it would be better if the % transmission readings are converted to % starch. This is achieved by adding fixed volumes of standard solutions of starch (that is, with a known concentration, such as 1%, 0.5%, 0.25% etc.) to dilute iodine solution and taking colorimeter readings. Plotting colorimeter readings against concentration of starch gives you a curve. This is a calibration curve — see Figure 10.

This curve is used to convert percentage transmission readings for the starch digestion over time experiment. Two values are shown being converted in Figure 10: 30% transmission (the value for the sample taken at 1.5 minutes) converts to 0.146% starch, while 50% transmission converts to 0.08% starch (the value for the sample at 2.5 minutes). Having completed this for all results, a graph of % starch against time may be plotted.

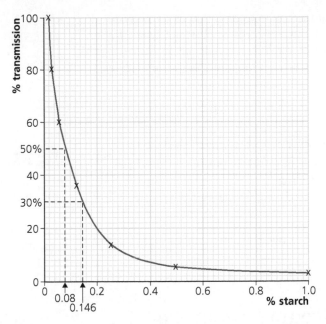

**Figure 10** A calibration curve for a colorimeter: % transmission against concentration of starch

## Interpretation and evaluation

The graph of % starch against time shows the course of the starch–amylase reaction — see Figure 11.

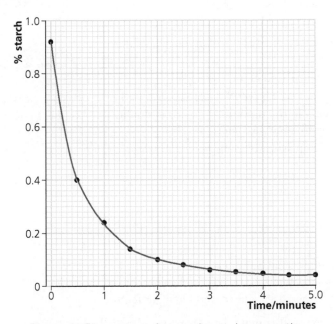

**Figure 11** The course of a starch–amylase reaction

The graph (an exponential decline curve) shows that as time progresses the amount of starch left in solution decreases rapidly initially and then the rate of digestion

**Knowledge check 16**

The product of starch digestion by amylase is maltose. What test could be used for the production of maltose?

slows down. Interpretation of this trend requires you to explain that as starch is digested there is less left for amylase to act on. In other words, substrate concentration decreases as the reaction progresses and so rate of reaction is reduced over time. The decrease in the rate of reaction may also be due to accumulating product molecules (maltose) interfering with the collision of substrate and enzyme molecules.

There are a number of important points to note in an evaluation of the experiment:
- the use of clean cuvettes
- the importance of the filter in maximising the range of readings taken
- the importance of standardising the colorimeter to 100% transmission using a 'blank' prior to taking each experimental reading
- the use of the calibration curve since the relationship between % transmission and % starch is not linear (Figure 10)

It is also noteworthy that the value for the sample taken initially contains less than 1% starch, representing a delay in sampling.

# Cells
# Cell fractionation and organelle isolation

To study the *function of cell organelles* they have first to be isolated. The technique for obtaining organelles is called **cell fractionation** and involves **homogenisation** and **centrifugation**. The procedure is summarised in Figure 36 in the AS Unit 1 Student Guide in this series (p. 34).

For example, to obtain a sample of chloroplasts these steps are followed:
- Cut fresh leaves are homogenised in cold isotonic buffer solution.
- This is filtered to remove cell debris.
- The supernatant is centrifuged (the first pellet is likely to contain nuclei, as they are the densest organelles).
- The supernatant is removed and further centrifuged at a faster speed.
- The second pellet should contain chloroplasts which are the next densest (and denser than mitochondria which will remain in the supernatant) — if in doubt, a sample of the pellet should be examined with a microscope.
- The supernatant is poured off and the chloroplasts are re-suspended in isotonic solution.

# Cell ultrastructure

You must recognise cell structures from photomicrographs, electron micrographs and drawings.

**Photomicrographs**, even at high magnification, show a limited number of cell structures: the nucleus and, in plants, cell walls, chloroplasts (coloured green) and starch grains. Staining may have been used to increase contrast and highlight features.

**Electron micrographs** have higher resolution so that more detail is provided of cell ultrastructure. The transmission electron micrograph (TEM) of part of a liver cell in Figure 12 shows many of the features of an animal cell. There is a high density of mitochondria (many in transverse section, t.s., some in longitudinal section, l.s.)

## Knowledge check 17

Describe the shape of a graph of products produced against time for an enzyme-controlled reaction.

## Exam tip

Notice that while cells are burst by homogenisation, the use of an isotonic solution prevents the *organelles* bursting by osmosis.

## Practical tip

When using a centrifuge, it is important that each tube is counterbalanced by a tube containing the same volume/mass of material.

indicating that liver cells have a high metabolic rate. The abundance of glycogen indicates that they act as 'energy' storage cells which may rapidly release glucose for respiratory purposes. The prominence of particular organelles in cells will reflect their function. For example, secretory cells, such as pancreatic cells, will have a large Golgi body and lots of vesicles.

**Exam tip**

Electron micrographs, even of the same tissue, will differ so you should expect that each requires interpretation. To identify mitochondria look for a surrounding envelope (double membrane) and the presence of cristae in the matrix.

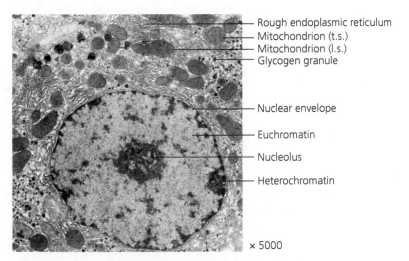

Rough endoplasmic reticulum
Mitochondrion (t.s.)
Mitochondrion (l.s.)
Glycogen granule

Nuclear envelope

Euchromatin

Nucleolus

Heterochromatin

× 5000

**Figure 12** Part of a liver cell (TEM)

An electron micrograph of a plant cell may additionally detail cell walls, with a middle lamella between adjacent cells, plasmodesmata, a permanent vacuole bound by a tonoplast, starch grains and the fine structure of chloroplasts.

Drawings of cell structures are idealised interpretations of micrographs. A drawing of the mitochondrion (l.s.) labelled in Figure 12 is shown in Figure 13. A scale bar is also shown.

**Exam tip**

A scanning electron micrograph (SEM) details the surface of structures, providing a 3-D effect.

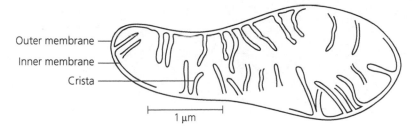

Outer membrane
Inner membrane
Crista

1 µm

Notice that the double membrane enveloping the mitochondrion and the structure of the cristae are incomplete in places since a drawing is a record of what the electron micrograph indicates.

**Figure 13** Drawing of a mitochondrion

# Calculating true size and magnification in micrographs

The length of the mitochondrion drawn in Figure 13 measures 77 mm, while the scale bar, representing 1 µm, measures 20 mm. The **true size** of the mitochondrion is calculated as:

$$\text{true size} = \frac{\text{measured size of image (mm)}}{\text{length of 1 μm on scale bar (mm)}}$$

$$= \frac{77}{20} \times 1\,\mu m = 3.85\,\mu m$$

The **magnification** is then calculated by the equation:

$$\text{magnification} = \frac{\text{measured size of image}}{\text{true size}} = \frac{77 \times 1000\,\mu m}{3.85\,\mu m}$$

$$= \times 20\,000$$

# Using an eyepiece graticule and stage micrometer to measure cell size

It is possible to measure the size of a cell, viewed with a compound light microscope, by using a graticule installed in the eyepiece. The **eyepiece graticule**, with a scale of 0 to 100, measures length in arbitrary units and needs to be calibrated. This is achieved using a **stage micrometer**. A stage micrometer is a slide with a very accurate scale etched onto it. Many micrometers have a scale of total length 1 mm divided into 100, so that each small division is 0.01 mm (10 μm).

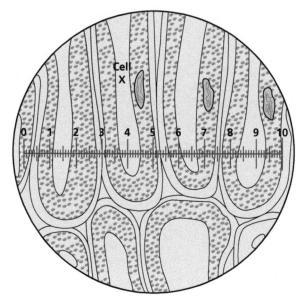

**Figure 14** Using an eyepiece graticule to measure the width of a cell

A specimen is viewed and the eyepiece turned so that the graticule scale lies over the object that you want to measure. Figure 14 shows the graticule scale adjusted to measure the width of a plant cell. The width of one cell is 23 small 'graticule units' (from 3.0 to 5.3 for cell X).

The specimen is replaced with the stage micrometer which is viewed using the same objective lens (*same magnification*) as used for viewing the specimen. The micrometer scale and eyepiece graticule scale are lined up, as shown in Figure 15.

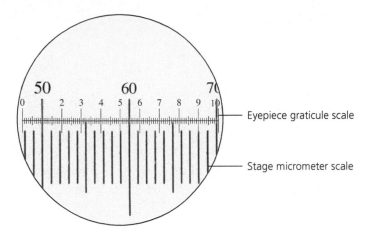

Eyepiece graticule scale

Stage micrometer scale

**Figure 15** Using a stage micrometer to calibrate an eyepiece graticule

You can see that the 50 mark on the stage micrometer is lined up with the 1.0 mark on the eyepiece graticule. Then, working towards the right to find another point at which the scales exactly match, you can see that there is good alignment of 68 on the stage micrometer and 9.0 on the eyepiece graticule. With a stage micrometer on which the smallest unit is 0.01 mm, the calibration is determined as:

80 small eyepiece graticule units = 18 stage micrometer units

$$= 18 \times 0.01 \, mm = 0.18 \, mm = 180 \, \mu m$$

So, 1 small eyepiece graticule unit $= \dfrac{180}{80} = 2.25 \, \mu m$

The true width of the plant cell shown in Figure 14 can then be calculated as:

Width of plant cell $= 23 \times 2.25 \, \mu m = 51.75 \, \mu m$

**Knowledge check 19**

Cells of onion epidermis were viewed with a microscope. An eyepiece graticule measured the length of an epidermal cell as 98 arbitrary units. Calibration with a stage micrometer showed that 90 units on the graticule equated to 24 units of 0.01 mm on the micrometer. Calculate the length of the cell in μm.

This calculation was undertaken using the high-power objective lens. If using a different (say, mid-power) objective lens, the eyepiece graticule units must be recalibrated using this lens.

# Drawing cells or cell sections

Making **drawings** is an important way of recording observations which are descriptive rather than quantitative. The purpose of a biological drawing is to make a clear record of what you have observed rather than an artistic impression.

**Practical tip**

To distinguish scales, simply rotate the eyepiece — the eyepiece graticule scale will also rotate but the stage micrometer is unaffected.

**Exam tip**

Reliability of your result will be increased by measuring a number of cells and calculating the mean.

Rules for drawing cells (or cell sections) include the following:
- Include a title and labels.
- Draw smooth (not sketchy), continuous lines, using a sharp pencil.
- Do not shade or add colour.
- Ensure that different parts of the drawing are in proportion to one another.
- Ensure drawing is large enough to show details clearly and state magnification.
- Label lines should not have an arrow head, should not cross and should be, as far as possible, horizontal.

Figure 16 shows a drawing of cells in (b), based on the photomicrograph of a leaf lower epidermis in (a). The photomicrograph shows air bubbles which are artefacts and not included in the drawing.

**Artefacts** These are structures in the image induced by the preparation of the specimen and not part of the original specimen.

**(a)**

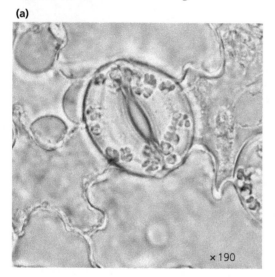

× 190

**(b)**

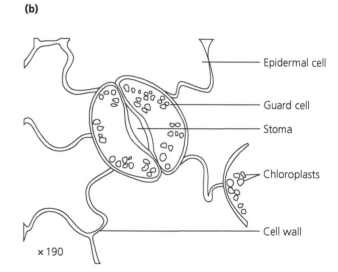

— Epidermal cell

— Guard cell

— Stoma

— Chloroplasts

— Cell wall

× 190

**Figure 16** (a) Photomicrograph of the lower epidermis of a leaf (*Tradescantia*) and (b) drawing based on (a)

# Using stains to aid observation

Structures, such as chloroplasts, are coloured and easy to see. To make other structures more obvious, coloured dyes may be used to stain them. As certain structures are highlighted by particular stains, contrast within the cell is increased.

Simple stains include the following:
- **Iodine** stains starch (in plant cells), showing up starch grains dark blue. It can also be used as a general stain to cause the nucleus and cell membranes to become more visible, e.g. to make plasmolysed cells more obvious.
- **Methylene blue**, a positively charged dye, stains the nucleus a deep blue colour since the DNA therein possesses a negative charge. The cytoplasm will show light blue in contrast. Methylene blue is known as a vital stain because it can enter cells without killing them.

Stains can be applied to a freshly prepared section on a microscope slide using the **irrigation technique**: the stain is applied to the edge of the coverslip and a piece of filter paper, placed at the opposite edge, is used to draw the liquid under the glass.

**Exam tip**

In an exam paper you may be asked to explain the advantage of using stains in the context of a colour photomicrograph provided in the paper.

# Plant cells and osmosis
## Measuring the average water potential of cells in a plant tissue

The **water potential** of plant tissue can be determined by the following principle. If a tissue shows no net gain or loss of water when immersed in a solution of known molarity, its water potential is equal to that of the external solution.

### Procedure

Using a core borer and razor blade, prepare six cylinders, at least 5 mm in diameter and 12 mm long, from the same large potato. Each potato cylinder is then cut into six discs about 2 mm thick. Groups of potato discs are surface-dried and weighed — this is the initial mass.

Each group of discs is immersed in water or one of a series of sucrose solutions of different molarity (0.2, 0.4, 0.6, 0.8, 1.0 M) and left for at least 2 hours to allow osmotic changes to occur. Then, each group of discs is removed, surface-dried (to remove *surplus* fluid) and weighed again — this is the final mass.

### Results

Percentage change in mass of each group of discs is calculated: change in mass multiplied by 100 and divided by the initial mass. If the experiment is repeated (for example, by other students) then mean percentage changes in mass may be calculated. This may be plotted against molarity of sucrose solution, or its water potential (using a conversion table). Figure 17 shows a graph of mean percentage change in mass against water potential of immersing sucrose solutions.

> **Practical tip**
>
> The conversion table shows the molarity of solutions and their corresponding water potentials, e.g. a 0.4 mol dm⁻³ solution has a water potential of –1120 kPa)

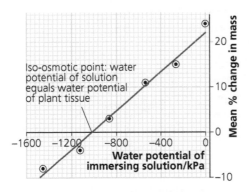

**Figure 17** Mean percentage change in mass plotted against water potential of immersing solution to enable the determination of the water potential of potato tissue

### Interpretation

The graph shows that where the sucrose solution has a very negative water potential (since it has a high molarity and is more concentrated) then the tissue loses mass. The tissue loses mass due to water being drawn out osmotically and so the tissue must have a higher water potential than the immersing sucrose solution. Where the sucrose solution has a less negative water potential (is dilute) the tissue gains mass (due to water moving in osmotically) and so the water potential of the tissue is lower than the immersing solution.

A line of best fit is drawn (since this best takes account of all the results in determining where the line crosses the horizontal axis). Where this line intersects the horizontal axis at zero percentage change in mass, you read off the water potential of the immersing sucrose solution. Since at this point $\psi_{tissue} = \psi_{external\ solution}$, this is also the water potential of the potato tissue. In Figure 17, this is determined as $-1020\,kPa$.

## Evaluation

There are important precautions to be observed in this experiment. The surface drying must be standardised: the pressure applied to the filter paper in drying the discs must be light and must be the same for each group of discs — you are only removing the *surplus* fluid. You must not squeeze the discs or they will *all* lose water. It is also important to work quickly to avoid loss of water from the discs through evaporation. You must be able to assume that the change in mass is due to osmotic loss or gain of water.

# Measuring the average solute potential of cells at incipient plasmolysis

Incipient plasmolysis is the stage when the plant cell is just about to become plasmolysed. At this point, the pressure potential is zero because the cell wall is not exerting any pressure on the protoplast. If the pressure potential ($\psi_p$) is zero, then the water potential of the cell ($\psi_{cell}$) is equal to its solute potential ($\psi_s$).

A solution with a solute potential which is equal to the solute potential of the cells will cause incipient plasmolysis. Therefore, to determine the solute potential of plant cells, you need to find the molarity of sucrose solution that causes incipient plasmolysis. Of course, it is not possible to recognise incipient plasmolysis in a single cell but, since plant cells will all differ slightly, the *average* solute potential can be determined when 50% of the cells are plasmolysed. Plasmolysed cells are recognised because the cytoplasm and sap vacuole both lose water and contract, so that the cell membrane pulls away from the cell wall (Figure 18).

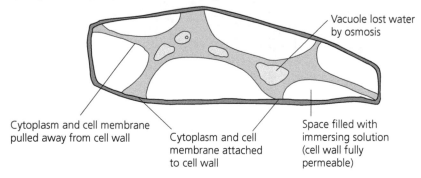

**Figure 18** A plasmolysed cell

## Procedure

In the experiment, squares of onion epidermis (or other suitable tissue) are placed in a series of sucrose solutions of different molarity (ranging from 0.3 to 0.8 M), left for some time (say, 20 minutes) and then viewed using a microscope under low power.

**Knowledge check 20**

In this experiment the water potential of the immersing sucrose solution was assumed to be equal to its solute potential. Explain why this assumption is acceptable.

The total number of cells within the field of view and the number identified as being plasmolysed are counted.

## Results

The percentage of cells plasmolysed (number plasmolysed divided by total viewed and multiplied by 100) is plotted against molarity or solute potential of sucrose solution, and a line of best fit drawn — see Figure 19.

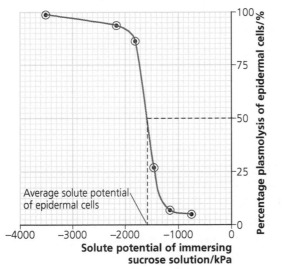

**Figure 19** Percentage plasmolysis plotted against solute potential of immersing sucrose solution to enable the determination of the average solute potential of the epidermal tissue

## Interpretation

The graph shows that most cells are plasmolysed in sucrose solutions of very negative solute potential (to which they lose water osmotically), while most cells are turgid in sucrose solutions of less negative solute potential. The average solute potential of the epidermis is determined at the point on the line at which there is 50% plasmolysis. In Figure 19 this is determined as −1570 kPa.

## Evaluation

The experiment relies on your ability to identify plasmolysed cells. This may be made easier by using red onion because of their coloured cell sap.

It is also important that you understand the theoretical basis of the experiment:

- After 20 minutes, the water potential of the tissue and immersing solution have equilibrated.
- The solution is under no pressure, so its water and solute potentials are the same.
- At 50% plasmolysis, the average pressure potential is zero, and so, on average, the water potential of the cells is equal to their solute potential — the solute potential of solution and tissue are the same.

**Knowledge check 21**

Suggest why the cells in the tissue do not all have the same solute potential. What might be the consequence of this?

# Mitosis and meiosis
## Preparing and staining a root tip squash to view stages of mitosis

In plants, cell division is restricted to specific areas called meristems. A good place to find dividing cells is just behind the root tip of a young plant. To be able to see chromosomes, it is necessary to:

- squash the root tip so that cells are spread out in a single layer
- stain the DNA so that chromosomes show up clearly when viewed through a microscope

### Procedure

1　Cut 5 mm off the end of a young growing root, e.g. lateral root of broad bean, or roots of onion or garlic.

2　Put the root tip into a small glass dish, containing dilute hydrochloric acid and acetic orcein stain. Warm this gently for a few minutes. Hydrochloric acid separates the cells, making it easier to squash the root at the next stage, while acetic orcein stains the DNA in the chromosomes red.

3　Now put the root tip onto the centre of a microscope slide. Cut off the end 2 mm (containing the root cap) and throw this away. Add two drops of acetic orcein onto the part still left (containing the meristem).

4　Put a coverslip over the root tip on the slide, cover it with filter paper and squash gently by pressing down on the coverslip. This process may need to be repeated to get the root tip cells spread out as much as possible, and ultimately flattened into a layer only one cell thick. Add a bit more stain if necessary. To intensify the staining, hold the slide over a Bunsen flame with your fingers for a few seconds — holding with your fingers will stop you letting it get too hot.

5　Examine the slide under a light microscope and identify any stages of mitosis.

### Results

You will find that most of the cells are in interphase (i.e. not undergoing mitosis) since cells remain in this phase for longer than any other (representing about 80% of the cell cycle). Of the cells in prophase, you may differentiate those in early prophase (chromosomes long and threadlike) and those in late prophase (chromosomes compact with chromatids obvious). Easiest to distinguish are the metaphase and anaphase stages, though these are short-lasting so that they are not common.

If you are asked to draw cells from your preparation, select a group of no more than five cells, including at least one that is undergoing mitosis. Remember to draw what you see, paying attention to cell and chromosome shape, and to the relative sizes of both.

> **Practical tip**
>
> Another common stain for DNA and chromosomes is toluidine blue — obviously staining chromosomes blue.

# Examining photomicrographs of mitosis and meiosis

You need to be able to recognise the stages of mitosis and meiosis from photomicrographs, and identify visible structures. Mitosis may be observed in areas of growth, such as the root tip of a plant or embryo of an animal, while meiosis may be seen in the reproductive organs of an animal or anther of a flowering plant.

Figure 20 shows a photomicrograph of separate stages of the cell cycle from an onion (*Allium* sp.) root tip squash.

**(a)** Interphase    **(b)** Prophase    **(c)** Metaphase    **(d)** Anaphase    **(e)** Telophase    **(f)** Cytokinesis

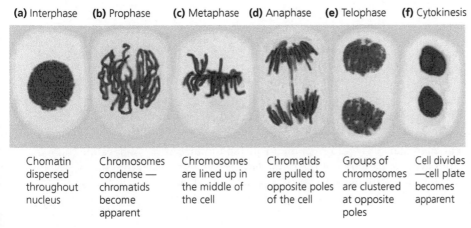

| Chomatin dispersed throughout nucleus | Chromosomes condense — chromatids become apparent | Chromosomes are lined up in the middle of the cell | Chromatids are pulled to opposite poles of the cell | Groups of chromosomes are clustered at opposite poles | Cell divides —cell plate becomes apparent |

**Figure 20** Stages of the cell cycle, including mitosis, from a root tip squash

You need to be able to distinguish meiosis from mitosis. Recognition of meiosis is possible because of the appearance of the chromosomes in certain stages of the first division:

- In prophase I, chromosomes appear as homologous pairs (bivalents) with a **tetrad of chromatids** between which **chiasmata** should be obvious — see Figure 21(a).
- In anaphase I, **pairs of chromatids** are moved apart (not just single chromatids) to opposite ends — see Figure 21(b).

**(a)**

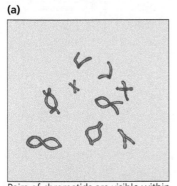

Pairs of chromatids are visible within bivalents (homologous chromosome pairs). Chromatids of paired homologous chromosomes are joined at chiasmata.

**(b)**

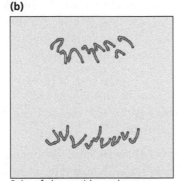

Pairs of chromatids can be seen separating (not individual chromatids) as homologous chromosomes separate.

**Figure 21** The appearance of chromosomes during (a) prophase I, and (b) anaphase I of meiosis

Chromosomes in the second division of meiosis behave as in mitosis, though the cell will be haploid — you can only determine this if you count the number of chromosomes and compare it with the diploid number for the species.

You should also be able to distinguish whether the dividing cell is animal or plant. If you can detect a cell wall then it is a plant cell. The presence of an **aster** indicates an animal cell — an aster, fibres radiating out away from the spindle at each pole, is produced from **centrioles** (not present in higher plants). The presence of a **cell plate** indicates a plant cell, while if you can observe a **cleavage furrow** then it is an animal cell.

# Tissues in the ileum and the leaf
## The ileum

You need to be able to recognise tissue layers and associated features from photomicrographs of the ileum: mucosa containing villi covered in a columnar epithelium within which are goblet cells, crypts of Lieberkühn at the base of which are Paneth cells, and each villus possessing a lacteal and blood capillaries; muscularis mucosa; submucosa; muscularis externa; and serosa. Examination of electron micrographs of columnar epithelial cells will reveal microvilli, and numerous mitochondria. You may be required to make drawings of the tissue layers of which the ileum is composed. A photomicrograph of the mucosa of the ileum is provided in the AS Unit 1 Student Guide in this series (pp. 92–94) along with students' attempted drawings.

## The mesophytic leaf

You need to be able to recognise tissues from photomicrographs of a mesophytic leaf: epidermis (covered by waxy cuticle), with lower epidermis containing guard cells surrounding stomata; palisade mesophyll; spongy mesophyll; and vascular tissue. You may be required to make drawings of a photomicrograph of a mesophytic leaf to indicate the different tissues.

In drawing a section of a leaf or the wall of the ileum, the emphasis must be on illustrating the tissue layers in a **tissue map** (sometimes called a **block diagram**). A tissue map outlines the different tissues. Individual cells are not generally drawn, except where required to show a few representative cells, such as guard cells in the lower epidermis of a leaf or goblet cells in the epithelium of the ileum.

Accepting that only tissue outlines are drawn, the rules for drawing are the same as for drawing cells (see page 20). You must draw the structures exactly as they appear in the photograph (or in a microscopic view of a prepared slide), and clearly represent the exact shape and proportion of the different tissues.

Annotated drawings may be required. Annotations are labels with more information than just the names of the structures. They give additional information about the features drawn such as functions, properties or observations. Figure 23 shows an annotated drawing in (b), based on the photomicrograph of a tea (*Camellia sinensis*) leaf in (a).

**Exam tip**

If identifying a particular stage of meiosis in an examination, from a photograph or drawing, make sure you specify the division (I or II) as well as the stage.

**Exam tip**

The muscularis mucosa is a thin layer of smooth muscle (spindle-shaped cells) separating the mucosa from the submucosa, and so allows you to identify the two layers.

**Mesophyte** A terrestrial plant from a temperate habitat (adapted to *neither* dry *nor* wet conditions).

**Exam tip**

In a leaf's vascular bundle, the xylem tissue always lies above the phloem.

**Exam tip**

When asked to draw a tissue map (or block diagram), resist producing an idealised diagram such as might appear in a textbook. It must represent the micrograph.

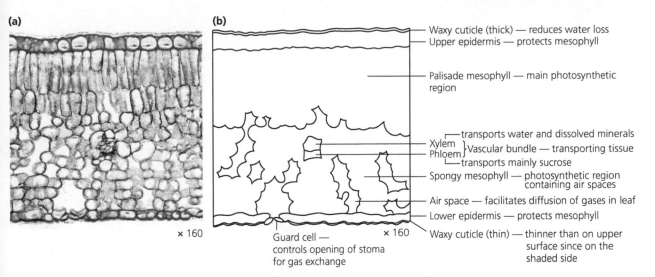

**Figure 22** (a) Photomicrograph of a tea (*Camellia sinensis*) leaf, t.s., and (b) an annotated drawing based on (a)

**Exam tip**

Annotated drawings are a useful revision exercise. This kind of drawing shows the labelled structures and also includes notes on functions and properties.

# ■ AS Unit 2 practical work

## Transport and exchange mechanisms
## Using a respirometer
### Measuring oxygen consumption

Respiring organisms absorb and use oxygen from their atmosphere, and produce and release carbon dioxide. In a closed vessel, consumption of oxygen by an organism will cause a reduction in pressure if carbon dioxide is chemically removed (absorbed by potassium hydroxide).

The respirometer (see Figure 23) has two identical closed vessels. One contains living organisms (e.g. germinating seeds) and the other acts as a *thermobarometer* — small changes in temperature or pressure cause air in this vessel to expand or contract, compensating for similar changes in the first vessel.

As oxygen is consumed, the level of the fluid in the manometer will rise up in the right-hand arm. The length of movement, over a set period of time (e.g. 10 minutes), represents a measure of the rate of oxygen consumption. If the diameter of the bore of the manometer tube is known, then the volume of oxygen consumed can be calculated (length $\times \pi r^2$). Alternatively, the syringe can be depressed to introduce a volume of air needed to equalise the levels in the manometer tube and the volume can be read off the syringe scale.

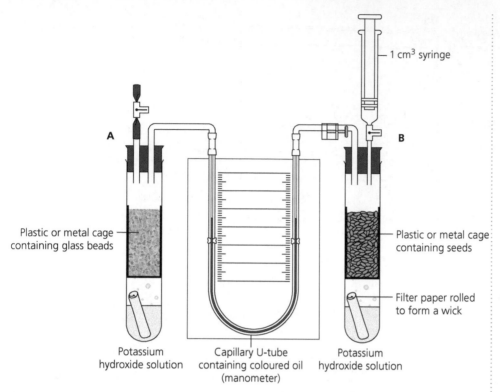

1 cm$^3$ syringe

A

B

Plastic or metal cage
containing glass beads

Plastic or metal cage
containing seeds

Filter paper rolled
to form a wick

Potassium
hydroxide solution

Capillary U-tube
containing coloured oil
(manometer)

Potassium
hydroxide solution

**Figure 23** A respirometer for measuring the rate of oxygen uptake by small organisms

The measure of oxygen consumption of a variety of living organisms can be considered, e.g. mung beans or maggots. Weighing the mass of the organisms allows the rate of oxygen consumption to be calculated as a volume per gram of tissue per unit time ($mm^3 g^{-1} min^{-1}$).

The respirometer is very sensitive to pressure changes and needs to be assembled carefully. This may be more readily demonstrated by your teacher. The apparatus must be tested for leakages at any of the joints. A decrease in the values obtained during the course of an experiment could be due to air entering from outside.

Since potassium hydroxide (KOH) is corrosive, lab coats and goggles must be worn. While there are no ethical issues if using plant material, animals must be handled with care and respect.

Simple respirometers lack the second tube acting as a thermobarometer, and so are easier to set up and use, though temperature or pressure changes during the course of the experiment cannot be compensated for.

## Measuring carbon dioxide production

The amount of carbon dioxide given out by respiring tissue is measured indirectly. If the potassium hydroxide is replaced with water, then carbon dioxide is not absorbed. In this case, any change in volume represents the *net difference* between carbon dioxide given out and oxygen taken up by the respiring tissue. If less carbon dioxide is given out than oxygen taken up, the fluid will move up, towards the respiring material, as volume is reduced. If the reverse happens and carbon dioxide production exceeds oxygen consumption, the fluid will move away from the respiring material. Given that oxygen consumption has been measured, then the amount of carbon dioxide produced by the respiring tissue can be calculated.

For example, measurements using the respirometer (with KOH present) may show that the respiring tissue consumed $20 \, mm^3$ of oxygen in 10 minutes. If the KOH is replaced with water, and the same respiring tissue shows a decrease in volume of $5 \, mm^3$ in 10 minutes (more $O_2$ consumed than $CO_2$ produced), then the amount of carbon dioxide produced is calculated as:

$$20 - 5 = 15 \, mm^3 \, 10 \, minute^{-1} \text{ (or } 1.5 \, mm^3 \, min^{-1})$$

# Tissues in plants
## Vascular tissues in the root and stem

You should be able to recognise, from photomicrographs, tissues involved in transport through the root and the stem:

- epidermis, cortex, endodermis, xylem and phloem within a stele in the root
- xylem and phloem within vascular bundles in the stem
- xylem containing vessels with different lignification patterns in protoxylem (annular or spiral) and metaxylem (reticulated or pitted)
- phloem containing sieve tube elements and companion cells

## Xerophytic and hydrophytic leaves

You need to be able to recognise tissues within the leaves of xerophytes and hydrophytes from photomicrographs and explain how these plants are structurally adapted to dry and wet habitats respectively. You should be prepared to draw tissue maps (block diagrams) of leaves.

# Heart, blood vessels and blood
## Heart dissection

You will be provided with the heart of a sheep or pig. Since it will have been obtained from a butcher, it may not be intact — for example, the atria are often removed and the major arteries may have been cut back. A helpful guide to dissecting a heart is found at www.nuffieldfoundation.org/practical-biology/looking-heart.

*Examination of external features*

- Distinguish between the ventral (front) and dorsal (back) sides of the heart — the ventral side is more rounded.
- Distinguish between arteries and veins: arteries have thick rubbery walls, while veins have much thinner walls. Both connect at the top (anterior end) of the heart, arteries towards the ventral side (aorta appearing to the heart's right side of the pulmonary artery and with a thicker wall) and veins towards the dorsal side (venae cavae on the right and the pulmonary veins on the heart's left side). See Figure 24.
- Observe the coronary arteries branching out over the surface and supplying the heart with blood.

**Practical tip**

Before dissecting a heart, you should familiarise yourself with its structure (see the AS Unit 2 Student Guide in this series, p. 28).

**Practical tip**

When the heart is placed dorsal side down on the dissecting board, the side that you view on your right is the heart's left side.

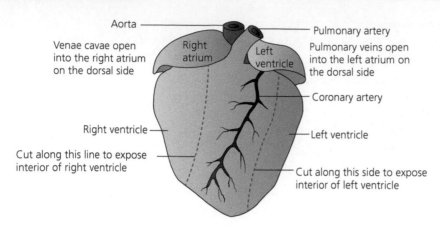

**Figure 24** The heart, viewed from its ventral side

## Dissection of the right and left ventricles

To examine the heart's interior, it is necessary to make two longitudinal cuts (with a scalpel) on the ventral side of the heart along the red broken lines shown in Figure 24.

Look into the **right ventricle**. Notice the three flaps of the valve (**tricuspid valve**) between the ventricle and the atrium. These flaps are attached by **chordae tendinae** to **papillary muscles** extending from the wall of the ventricle. A probe (a coloured straw is effective) can be inserted into the **right atrium** and passed through the **venae cavae** (confirming their identification). Examine the entrance to the **pulmonary artery** and identify the **semilunar valves** — these 'pockets' can be further investigated by opening up the artery (with dissecting scissors). If the artery is not cut open, a probe can be passed through to identify the emerging pulmonary artery. Similarly, the atrium can be cut open to allow you to compare the thinness of its wall to that of the ventricle.

Now look into the **left ventricle**. Notice that the ventricle and atrium are separated by two flaps (**bicuspid valve**). Again, identify the chordae tendinae and papillary muscles. Pass a probe through the **left atrium** to identify the **pulmonary veins**. Notice the semilunar valves at the entrance to the **aorta**. Pass a probe through to confirm the emerging aorta.

Compare the cut walls of the two ventricles. Notice that the wall of the left ventricle is much thicker (generating a higher pressure in the functioning heart). The wall separating the two ventricles is the **interventricular septum**.

## Arteries and veins

You need to be able to interpret photomicrographs of arteries and veins, distinguishing them and recognising different tissues. Arteries have a smaller lumen but thicker wall. The walls of both have three layers, with similar inner and outer layers: they have an inner layer composed of a simple squamous endothelium, and an outer layer containing protective fibrous tissue. However, the middle layer, composed of smooth muscle and elastic tissue, is much thicker in arteries (accounting for their thick wall).

Capillaries have a simple wall composed of a single layer of squamous endothelium.

You must also be able to identify the major blood vessels of the thorax and abdomen.

## Blood cells

You need to be able to recognise different blood cells from photomicrographs (or microscopic slides) of stained blood smears. Red blood cells, polymorphs, monocytes, lymphocytes and platelets are illustrated in Figure 25.

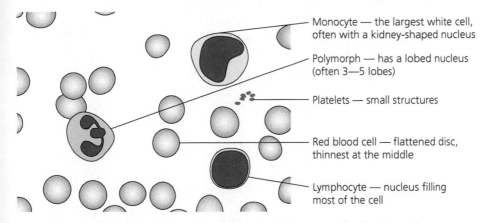

Monocyte — the largest white cell, often with a kidney-shaped nucleus

Polymorph — has a lobed nucleus (often 3—5 lobes)

Platelets — small structures

Red blood cell — flattened disc, thinnest at the middle

Lymphocyte — nucleus filling most of the cell

**Figure 25** Illustration of blood cells in a stained blood smear

# Ecological techniques

Ecological investigations are usually aimed at finding out something along the following lines:

- What types of organisms are found here?
- How abundant are they?
- Where are they most abundant within the area?
- What abiotic (environmental) factors might be responsible for the distribution?

## Sampling procedures

When investigating populations of organisms in a habitat it is too time-consuming and not practical to study every individual so you need to study a **representative sample** — a sample that is typical of the whole population. The type of sampling procedure will depend on what you are trying to achieve.

## Random sampling

To study a plant or animal population in a reasonably uniform habitat, such as a meadow, you need to use a sampling device. For plants and sessile (non-moving) or slowly moving animals (such as limpets) this is often a quadrat. This device is positioned randomly throughout the area. In random sampling every individual has an equal chance of being included in the sample, so there is no bias. This avoids sampling only in the middle of the area, or selecting samples where you think the plant, or animal, might live.

How many samples should be taken? Obviously, the more samples you take then the more reliable will be your estimate for the population. Generally in ecology, between 20 and 60 random samples are taken. This amount of replication is necessary because of the highly variable nature of any habitat. The actual amount of replication will

**Sampling** Taking results from part of a population or small areas within a habitat.

**Random sampling** Each individual or area selected is not chosen by the investigator, to ensure results are representative and unbiased.

depend on the amount of information assessed at each sample site: estimating the abundance of all species present is time-consuming and so 20 samples will suffice; determining the presence of a single species should not be too time-consuming, making possible a much higher level of replication.

How is a random sample achieved? Figure 26 shows how to place quadrat frames at random. You should:

- consider the area to be sampled as a grid with the bottom and left sides as axes — tape measures should be placed along these sides
- use a random number generator on a calculator to produce a pair of coordinates, e.g. X5, Y4
- place a quadrat with its bottom left-hand corner on the intersection of the coordinates
- estimate the abundance of the species inside the quadrat
- repeat the random placement of quadrats an appropriate number of times

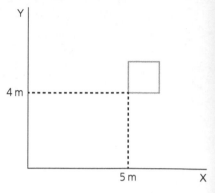

**Figure 26** Placing a quadrat at random

# Transect sampling

**Transect sampling** is used in the situation where there is a transition in the environment and zones of different species are present. These zones of different species types represent a **zonation**. Examples of **environmental gradients** may be observed in the distribution of:

- seaweeds and animals on a rocky shore
- plant species at the edge of a lake
- animal and plant species from a meadow into woodland

## Line transect

A **line transect** is set up across the areas where there are clear environmental gradients. A line (or tape) is stretched along the environmental gradient and any plant species touching the line are recorded, either continuously or at marked intervals (every 0.5 m or 1 m). The species can be presented diagrammatically as being either **present or absent** along the length of the transect.

While information about species presence along the transect can be quickly obtained, the data produced by a line transect are somewhat limited. Data collection may be improved by repeating line transects several times through the habitat.

## Belt transect

A **belt transect** will supply more data than a line transect. It will give data on the *abundance* of individual species at different points along the line of transect. In a belt transect, the transect line is laid out along the environmental gradient and a quadrat is placed on the first marked point on the line. The plants and/or animals inside the quadrat are then identified and their abundance is estimated. Animals can be counted, while it is usual to estimate the percentage cover of plant species.

Quadrats are sampled all the way along the transect. If the line of transect is relatively short, say 20 m or less, then 1 × 1 m quadrats might be placed **contiguously** (end-to-end). If the length of the transect is much longer then an **interrupted belt transect**

### Knowledge check 24

What can be deduced from line transect data? What cannot be deduced from line transect data?

is used: the quadrat is positioned at intervals, such as every 2, 5 or 10 m, depending on the overall length to be sampled — see Figure 27.

**Practical tip**

Remember that a belt transect would be interrupted if the distance is more than approximately 20 metres.

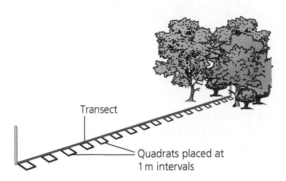

Transect

Quadrats placed at 1 m intervals

**Figure 27** An interrupted belt transect

Data from a belt transect can be used to construct a kite diagram. Figure 28 shows the distribution of seaweeds and some animals on a rocky shore. For each species, abundance is shown symmetrically round a value of zero. The data were obtained by placing a quadrat at the low water mark and sampling every other metre up the shore.

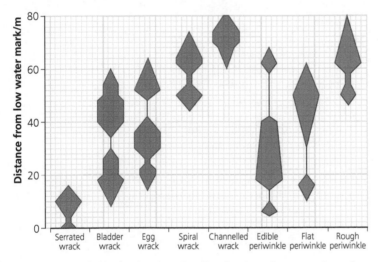

**Figure 28** A kite diagram showing the distribution of seaweeds and some animals on a rocky shore

# Sampling devices
## Quadrats

A **quadrat** (or frame quadrat) is an area in which *plants or non-mobile animals* can be sampled. Quadrats are usually square: ½ × ½ m and 1 × 1 m are common but larger quadrats can be constructed. Generally, smaller quadrats are used to sample smaller plants and those that are common in the area sampled, while a larger quadrat size is required for larger plants and for less common plants. Also, less common species require a greater number of samples (since the species might be expected to be absent from many of the quadrats).

To *estimate the abundance of a species* in an area:

● determine the abundance in a number of quadrats (say thirty ½ × ½ m quadrats)
● calculate the mean abundance per quadrat (for a ½ × ½ m quadrat this is 0.25 m²)
● estimate the total area within which sampling has taken place
● determine the overall abundance by dividing the mean abundance in a quadrat by the quadrat area and multiplying this by the total area

## Pin frames

A **pin frame** (**point quadrat**) consists of a frame containing a horizontal bar with holes (usually ten) through which long pins can be pushed to reach the ground — see Figure 29. They are used for plant sampling in *short grassland*.

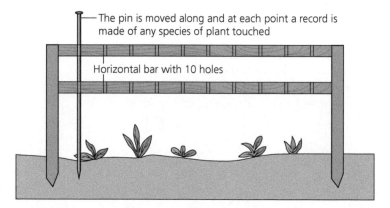

The pin is moved along and at each point a record is made of any species of plant touched

Horizontal bar with 10 holes

**Figure 29** A pin frame (point quadrat)

The frame is positioned (either randomly or along a transect) and pins are lowered onto the vegetation. Each species of plant touched by a pin is recorded. For each species, the number of recorded 'touches' (or 'hits') divided by the total number of pins lowered is calculated as percentage cover (the proportion of the ground area covered). For example, if a species is touched by 3 out of the 10 pins in a frame, the percentage cover is 30% at that sampling location.

## Pitfall traps

A **pitfall trap** is used to capture arthropods (insects, spiders etc.) *walking over the soil surface*. A jar is inserted in the soil so that its rim is flush with the surface, and roofed to prevent it flooding with rainwater (see Figure 30). Passing arthropods, such as beetles, fall into the jar and cannot escape. The jar should be examined at frequent intervals (to reduce the likelihood of carnivores eating other arthropods) and living organisms removed for identification and counting.

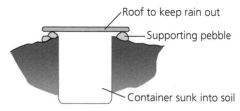

Roof to keep rain out

Supporting pebble

Container sunk into soil

**Figure 30** A pitfall trap, used to sample arthropods moving over the soil surface

Pitfall traps may be randomly arranged in a habitat to determine the types of arthropods there or to estimate the relative density of a particular species of beetle; or they may be arranged along a transect through habitat boundaries, such as through the edge of woodland leading into a meadow.

## Sweep nets and pooters

A **sweep net** is a net used for sampling arthropods *from tall grassland*. A standard number of sweeps, say 10 sweeps from one side to the other, is undertaken. The arthropods collected are trapped in a **pooter** (see Figure 31). The organisms trapped in the sample are identified and counted. This is a particularly useful technique for determining the relative abundance of an insect inhabiting a meadow from spring through to autumn. The data in Figure 34 were obtained using a sweep net and pooter.

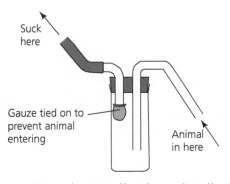

Suck here

Gauze tied on to prevent animal entering

Animal in here

**Figure 31** A pooter, used to suck up small arthropods collected in a sweep net

# Estimation of species abundance

There are different measures of abundance, e.g. density, percentage cover and frequency.

## Density

Density is simply the *number of organisms per unit area*. This is appropriate for determining the abundance of animals, such as limpets (on a rocky shore), which move very slowly and are not hidden, or for plant species in which individuals are easily recognised, such as orchids. The number of individuals of a species within a quadrat is counted and, after repeating this in different positions, the results can be expressed as the average number of individuals per unit area — the density.

## Percentage cover

Often it is difficult, with plants, to count individuals. For example, do the blades of grass belong to one individual or several? Also, plants often differ greatly in size and so counting numbers is not necessarily a good estimate of abundance. For example, on the rocky shore there may be small and very large specimens of the seaweed eggwrack (*Ascophyllum nodosum*). To get around this, you should estimate the **percentage cover** of a species.

Using a quadrat, you estimate the portion of the area covered by the plant or seaweed species under investigation. If it covers a quarter of the quadrat then the percentage cover is 25%. A quadrat subdivided by a grid into smaller squares will

**Knowledge check 23**

A lawn, 9 metres by 20 metres, was sampled using thirty ½ × ½ m quadrats to estimate the percentage cover of creeping buttercup. The average percentage cover per quadrat was found to be 2%. Determine (a) the percentage cover of the whole lawn, and (b) the total area of the lawn covered by creeping buttercup.

**Practical tip**

In print, scientific names are given in italics (e.g. *Ascophyllum nodosum*). In your handwritten work, you should underline them.

help in estimating percentage cover. In a 5 by 5 grid, each square is 1/25 or 4% of the total area (see Figure 32). If the species covers 4 whole squares and parts of other squares that are equivalent to 3 squares, then a total of 7 squares is estimated, giving 7 × 4% = 28%.

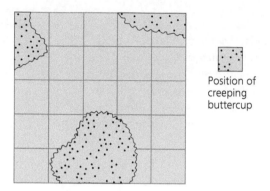

Position of creeping buttercup

**Figure 32** The cover of creeping buttercup (*Ranunculus repens*) in a quadrat

**Knowledge check 26**

Determine the percentage cover of creeping buttercup in the quadrat shown in Figure 32.

## Frequency

The determination of frequency uses **presence–absence data**. Is the plant present or is it absent? Data may be collected by sampling along a line transect or by looking for presence in a series of quadrats. Frequency is calculated as *the number of times that a particular plant species is present at sampling points divided by the total number of sampling points*. Using quadrats, this is the percentage of quadrats that included each species.

Frequency is not a measure of the abundance at any sampling point and so provides rather limited data. Its use is in areas where there is a lot of data to obtain and determining presence or absence allows this to be completed quickly.

## Influence of environmental factors on the distribution of organisms

At each sample point, the environmental factors that might influence the distribution of organisms can be monitored. For example, nettles (*Urtica dioica*) appear to grow most densely in soils with high phosphate levels, while the meadow buttercup (*Ranunculus acris*) tends to be more frequently found in damp meadows.

**Abiotic factors** might include the following:

- **Light intensity** — this can be measured using a light meter.
- **Soil water content** — the soil sample is weighed, dried in an oven to constant mass at 105°C and weighed again. The decrease in mass (the amount of water evaporated) divided by the dry mass × 100% represents the percentage water content of the sample.
- **Soil organic content** — a known mass of oven-dried soil is placed in a crucible, the organic matter is burnt off and the soil weighed again. The decrease in mass (the organic matter) divided by the initial dry mass × 100% represents the percentage organic content.
- **Soil temperature** — this can be measured using special soil thermometers.

**Exam tip**

Remember that in determining Simpson's index (see the AS Unit 2 Student Guide in this series, p. 51), you must be consistent in using either density or percentage cover or frequency.

- **Soil pH** — this can be measured using a soil testing kit (using pH indicators).
- **Nutrient availability in the soil** — certain nutrients can be measured using a soil testing kit.

You can expect to undertake **fieldwork** to collect data on populations of plants and/or animals in their environment. You are expected to know about sampling, sampling devices and measures of abundance. You are also expected to be able to *explain* your results or, at least, make *reasoned suggestions*, and to *evaluate* the procedures used and the data collected. When evaluating ecological tasks, expect to comment on the following problems:

- Sampling may be biased so that the results are not representative.
- Too few samples may be taken so that results are not reliable.
- Estimating the abundance of different species may be difficult so that errors decrease the accuracy of the results.
- Identifying different species may be difficult so that misidentification reduces the validity of the conclusions.

# ■ Dealing with data: mathematical skills

## Units of measurement

The units that you meet will be based on the SI system. Table 3 shows those commonly used.

**Table 3** SI units and their derivatives

| Measure | Base unit | Symbol | Derived units commonly used |
|---------|-----------|--------|-----------------------------|
| Length | metre | m | $1\,m = 10^3$ millimetre (mm)<br>$1\,mm = 10^3$ micrometre (µm)<br>$1\,µm = 10^3$ nanometre (nm) |
| Area | square metre | $m^2$ | $1\,m^2 = 10^4$ cubic centimetre ($cm^3$)<br>$= 10^6\,mm^2$ |
| Volume | cubic metre<br>cubic decimetre | $m^3$<br>$dm^3$ | $1\,m^3 = 10^9\,mm^3$<br>$1\,dm^3 = 10^3\,cm^3$ |
| Mass | kilogram | kg | $1\,kg = 10^3$ gram (g)<br>$1\,g = 10^3$ milligram (mg) |
| Time | second | s | $60\,s = 1$ minute (min)<br>$60\,min = 1$ hour (h) |
| Temperature | degrees Celsius | °C | |
| Pressure | pascal | Pa | $10^3\,Pa = 1$ kilopascal (kPa) |
| Energy | joule | J | $10^3\,J = 1$ kilojoule (kJ) |
| Amount of substance | mole | mol | $1\,mol = 10^3$ millimole (mmol) |

### Practical tip

The older units of litre (l) and millilitre (ml) will be seen on some laboratory glassware: $1\,l = 1\,dm^3$ while $1\,ml = 1\,cm^3$.

### Practical tip

Avoid using centimetres for measuring length, for example of a scale bar on a micrograph. Use millimetres and you are less likely to make a mistake if you need to convert them.

### Knowledge check 27

Calculate the number of $mm^3$ in $1\,cm^3$.

### Practical tip

Avoid using mixed units such as minutes and seconds. Convert the whole time to seconds.

Whether the base or derived unit is used will be dictated by the situation. For example, it would not be sensible to record the size of a cell organelle in metres, or the time period for an investigation lasting weeks or months in seconds.

## Concentration

The concentration of a solution is the amount of a substance in a given volume of the solvent. It is correctly expressed as moles per cubic decimetre ($mol\,dm^{-3}$). However, it is easier to make up percentage solutions where a 1% solution has 10 g of the substance dissolved in $1\,dm^3$ of the solvent.

## Rate

Rates involve combining a measure with time, for example in a rate such as how fast oxygen is being produced ($mm^3\,min^{-1}$) or how fast an animal is moving ($mm\,s^{-1}$).

For some rates, the first measure is a count such as the number of beats in a heart rate: $beats\,min^{-1}$.

Sometimes two measures are combined with time, such as when giving figures for gross and net production ($kJ\,m^{-2}\,day^{-1}$ or $kJ\,m^{-3}\,day^{-1}$ depending on whether it is a terrestrial or aquatic habitat respectively).

## Expressing numbers
### Decimal places, rounding and significant places

When carrying out calculations the answer will often include a number of **decimal places**. There is no point in quoting all the decimal places, since this would give a misleading level of accuracy. The rule is not to make the result more accurate than the measurements that you are using.

For example, the mass of a potato cylinder, using a digital balance that has an accuracy of 0.01 g, may be recorded as 3.67 g. However, when calculating the mean of a number of potato cylinders you may arrive at an answer of 3.49657. So the figure is **rounded**, in this case to two decimal places by considering the number at the third decimal place, which is more than 5 so that the mean mass is recorded as 3.50 g. Notice that the calculated mean has the same number of **significant figures** as the individual measurements, that is three (since the zero at the second decimal place is significant).

In a calculation involving a division, if the numerator has three significant figures but the denominator has five significant figures, then the answer should only have three significant figures. This is the case when calculating the true size of a cell organelle from the magnified image on an electron micrograph. For example, a chloroplast measuring 115 mm in an electron micrograph with magnification × 15 000 would result in a calculation of 115 ÷ 15 000 × 1000 (conversion to µm) giving a figure of 7.666667. Since the chloroplast was measured in millimetres to three significant figures, the actual size of the chloroplast should be recorded as 7.67 µm.

> **Exam tip**
>
> When presenting the result of a calculation, say of a mean, do not simply write down all the decimal places — you will not gain the mark. Round appropriately.

## Ordinary and standard form

Numbers which in **ordinary form** are very large or very small are often presented in **standard form**. This will be in the form $a \times 10^b$, where $a$ is the number between 1 and 10 and $b$ is a power of 10. For example, an electron micrograph magnification of 21 000 is written as $2.1 \times 10^4$. For very small numbers of less than 1, the power of 10 becomes negative, meaning the same as 'divided by'. For example, a concentration of $0.0050\,\text{mol}\,\text{dm}^{-3}$ is converted as $5.0 \times 10^{-3}\,\text{mol}\,\text{dm}^{-3}$.

## Estimating calculated values

When using a calculator it is very easy to make a slip when punching in the numbers. You should estimate an answer to check that the calculated value is appropriate. You can do this in your head or by jotting down numbers on paper and doing a calculation with numbers that are rounded.

## Calculating surface areas and volumes

You should know how to calculate the surface areas and volumes of regular shapes such as cubes, cylinders and spheres. The formulae for these calculations and examples of their uses are shown in Table 4.

**Table 4** Formulae for calculating surface area and volume of regular shapes, and examples of their use

| Shape | Surface area | Volume | Example of use |
|-------|--------------|--------|----------------|
| Cube | $6l^2$ | $l^3$ | To illustrate decreases in surface-area-to-volume ratios with increasing size |
| Cylinder | $2\pi rh$ | $\pi r^2 h$ | To determine the volume of oxygen used in a respirometer; to estimate the surface area and volume of an earthworm |
| Sphere | $4\pi r^2$ | $\frac{4}{3}\pi r^3$ | To estimate the surface area and volume of cells which are nearly spherical, e.g. yeast, some bacteria and many protoctists |

The symbols used are: $l$ = length of cube; $\pi$ = 3.14; $r$ = radius of a circle or sphere; $h$ = height of cylinder.

Surface areas and volumes are important with respect to oxygen uptake and the respiratory needs of an organism.

## Using ratios and percentages
### Ratios

Ratios are used to indicate the relationship between one attribute and another. Examples of their use in biology include the following:

- Surface-area-to-volume ratios are used to demonstrate why larger organisms need specialised exchange surfaces.
- Ratios of DNA base pairs, A:T and G:C, are 1:1, a finding which provided evidence for the double-strand nature of DNA structure.
- Phenotypic ratios in genetic crosses can indicate the nature of a character's inheritance. For example, a 3:1 ratio indicates that the parents are heterozygous and that one character (the more common) is dominant.

**Knowledge check 28**

A magnification is calculated as 25040. How would this be presented in standard form and to 3 significant places?

**Exam tip**

Given a calculation of $9.7 \times 597.6 \div 105.6$, you would estimate the answer as $10 \times 600 \div 100$, and so the answer would be in the range of 60.

**Exam tip**

You have already learnt about the relationship between cube size and the surface-area-to-volume ratio in AS Unit 2 (see p. 7 of the AS Unit 2 Student Guide in this series).

**Exam tip**

If you need to calculate the surface area and volume of a cylinder or a sphere, the formulae will be given to you.

## Percentages

Calculating percentages is a way of making valid comparisons between items when the totals are different. For example, if you wanted to compare the incidence of heart disease in Scotland with that in England, you could not directly compare the number of people with heart disease, because the population of Scotland is much smaller than that of England. Instead, you would need to calculate the percentage of people with heart disease in each country.

Percentage is calculated as:

$$\% = \frac{\text{number with feature under study}}{\text{total number}} \times 100$$

## Percentage change

If you are recording measurements in an experiment where the starting values are all different, you can calculate percentage changes to make sure your results are comparable. For example, when investigating the increase or decrease in mass of potato cylinders in different sucrose concentrations, the starting mass of the cylinders will be slightly different, even if you try and cut them to the same size. So an increase or decrease in mass can be expressed as a percentage of the original mass, and all the results become directly comparable.

Percentage change is calculated as:

$$\% \text{ change} = \frac{\text{difference between final and original measurement}}{\text{original measurement}} \times 100$$

## Calculating averages

There are three measures of average: mean, median and mode.

### Mean

The mean is calculated arithmetically by using the formula:

$$\bar{x} = \frac{\Sigma x_i}{n}$$

where $\bar{x}$ = mean, $\Sigma$ = the sum of, $x_i$ = any value for the measurements made, and $n$ = the total number of values.

The mean is the best form of average to calculate if the data shows a **normal distribution** (a bell-shaped curve).

### Median

If the data is not normally distributed but **skewed** (with the majority of values on one side of the distribution), the median is the best average to use. The median is in the middle when the values are ranked in order of magnitude. If there is an odd number of values, the median is the middle value; if there is an even number then the median is the mean of the two central values.

### Mode

The mode is the value that occurs most frequently in a set of data. The mode is useful in describing a distribution with two peaks, called a **bimodal distribution**, where two values occur most frequently.

# Graphs

A graph is an illustration of how variables relate to one another. It is basically a 'picture' of the results. It allows the trend or pattern in the data to be more easily seen. Graphs are an aid to understanding and interpretation.

Different types of graph are available depending on the nature of the variables and on the type of data. There are *four main types of graph* that you might construct:

- line graphs
- histograms
- bar graphs (bar charts)
- scatter graphs (scattergrams)

Different graphical techniques are summarised in Table 5.

**Exam tip**

Graphs are drawn once data is presented in a table. You have learned about constructing tables in AS Unit 2 (see pp. 50–51 in the AS Unit 2 Student Guide in this series).

**Table 5** The appropriate use of different graphical techniques

| Graphical technique | Independent variable (IV) | Dependent variable (DV) | Appropriate use | Example |
|---|---|---|---|---|
| Line graph | Continuous data | Continuous data | To determine the nature of a causal link between the IV and the DV | The effect of pH on the rate of reaction of two protease enzymes |
| Histogram | Continuous data that has been subdivided into classes | Quantitative data (which may be discrete — frequencies) | To illustrate the frequency distribution of the IV in a population | A frequency distribution of leaf width of wild garlic (*Allium ursinum*) leaves |
| Bar graph (bar chart) | Qualitative data — different attributes or categories | Continuous data | To illustrate the effect of different attributes on the DV | The effect of different colours of light on the rate of photosynthesis |
| Scatter graph (scattergram) | Both variables have continuous data though there may not be an obvious independent variable, i.e. they may be interdependent | | To determine the relationship between two variables where there is not necessarily a causal link | The relationship between soil pH and the diameter of heather (*Calluna vulgaris*) stems |

Irrespective of the type of graph, there are guidelines concerning its construction. These are summarised in Table 6.

**Table 6** General guidelines for the construction of graphs

| Feature | Description |
|---|---|
| Caption or title | This must include the independent variable (e.g. The effect of ... on), the dependent variable (the process being affected) and the biological material being investigated. |
| Axes right way round | The independent variable (if known) must be presented on the *x*-axis. |
| Appropriate scaling | Appropriate scales for both axes should be devised to make maximum use of the graph paper — the graph should fit on the paper yet be sufficiently large to allow a more accurate plotting of results and facilitate interpretation. The data should be critically examined to establish whether it is necessary to start the scale at zero. |
| Axes with labels and units of measurement | Each axis should be clearly labelled. Each label should be followed by a solidus (slash) and then the unit of measurement. The axes may be offset to prevent data points being obscured by the scale line (see Figure 34 on p. 43). |
| Accurate plotting | All points or bars (columns) should be plotted accurately. It is best that a sharp pencil is used. If two sets of data are plotted on the same axes then each set needs to be labelled or a key added. |

The caption and axis labels are important since someone should be able to look at the graph and know exactly what it is showing without any further explanation.

## Line graphs

### Joining the points

Points may be notated with a saltire cross (×) or an encircled dot (⊙ — though it is possible to have a triangle or square around the dot, or have the symbol filled in). Both may be used when two or more sets of data are to be plotted on the same set of axes. It is only for these points that measurements have been taken. It is not actually known what is happening between the points, so why join them up? Basically, there are two reasons for joining up the points:

- It makes it easier to see the trend or relationship between the two variables.
- It may be important to determine what is happening between points — a process known as *interpolation* — in order to make a calculation.

Once the points are plotted a decision must be made as to whether to draw a line of best fit (or trend line) — either a straight line or, if theoretical considerations allow or the general trend of the points indicate, a smooth curve — or a series of short straight lines joining the points. Conventions for A-level biology have been established by the Institute of Biology who provide the following guidance:

- *A smooth curve can be drawn to represent the relationship when there are sufficient data points to be confident in the relationship or, because from theory, there is good reason to believe that the intermediate values fall on the curve, e.g. the course of an enzyme-controlled reaction.*
- *When there is insufficient data to be able to confidently interpolate the relationship, straight lines joining the points should be drawn, thus indicating uncertainty about the intermediate values, e.g. numbers of ground beetles found at set distances from a hedge.*

In most cases, it is better to join the points with straight lines because you do not know how the values between the recorded points may vary. Use a ruler to produce the straight lines.

Figure 33 shows line graphs of data for the production of gas by *Elodea* (an aquatic plant) at different relative light intensities. Short straight lines are the preferred method of joining the points and, while a smooth curve might be acceptable, it is less easy to achieve — see Figure 33(a). What is not acceptable is an undulating curve or a smooth curve which is not best fit — see Figure 33(b). Note also that it is not appropriate in this case to continue the line to the 0,0 coordinates ('origin').

Figure 34 shows the results of sampling two species of insect with a sweep net during one summer in a meadow at Kinnego Bay. Notice that drawing smooth curves would be totally inappropriate. The data is prone to experimental error and so there is uncertainty about the results.

**Practical tip**

When drawing line graphs with multiple lines, use a different colour or type of line for each variable to ensure that your grap is clear and readable.

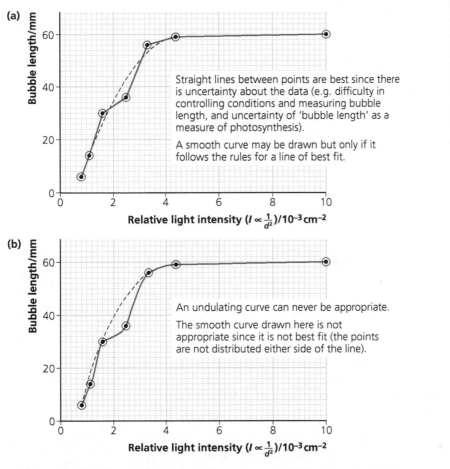

**(a)** Straight lines between points are best since there is uncertainty about the data (e.g. difficulty in controlling conditions and measuring bubble length, and uncertainty of 'bubble length' as a measure of photosynthesis).

A smooth curve may be drawn but only if it follows the rules for a line of best fit.

**(b)** An undulating curve can never be appropriate.

The smooth curve drawn here is not appropriate since it is not best fit (the points are not distributed either side of the line).

**Figure 33** The effect of light intensity on the production of gas by *Elodea*: (a) CORRECT (i.e. short straight lines), (b) INCORRECT (i.e. undulating curve)

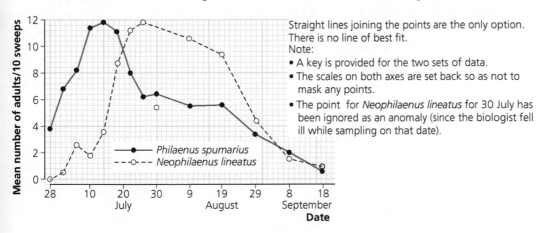

Straight lines joining the points are the only option. There is no line of best fit.
Note:
- A key is provided for the two sets of data.
- The scales on both axes are set back so as not to mask any points.
- The point for *Neophilaenus lineatus* for 30 July has been ignored as an anomaly (since the biologist fell ill while sampling on that date).

**Figure 34** The relative frequencies of the adults of two spittlebug species (*Philaenus spumarius* and *Neophilaenus lineatus*) sampled with a sweep net during one summer in a meadow

## Anomalous results

After plotting the points you can decide if any are anomalous. Ask yourself the question: do they fit the trend? What should you do if you spot an anomaly? Unless you can account for the anomalous result (possibly due to some fault in the procedure or equipment) you are stuck with it. If you can explain the anomaly then the datum point may be ignored. An instance like that is shown in Figure 34, where the biologist fell ill and developed a high temperature on the day of sampling (30 July).

> **Anomalous result**
> Result which does not fit the trend due to experimental error.

## Lines of best fit

There is one situation where you must draw a straight line or a smooth curve (a trend line). It may be important to interpolate the data points in order to make a calculation. If interpolation between two points is required, accuracy is improved by drawing a line of best fit so that all points are considered and not just the two points between which a calculation is to be made. There are a number of situations in which you have to make calculations from a graph. These include:

- constructing a calibration curve for colorimeter reading (% transmission) against percentage starch (using iodine to produce a blue-black solution) — a *smooth curve* is required (see Figure 10 on p. 15)
- plotting percentage change in mass of potato tuber against water potential of the bathing sucrose solution (or its molarity) and determining the water potential of the potato tuber where the line of best fit (a *straight line*) intersects the x-axis with no net change in mass (see Figure 17 on p. 21)
- plotting percentage plasmolysis against water potential of the bathing sucrose solution (or its molarity) and determining the average solute potential of the epidermal tissue from the line of best fit (most likely a *sigmoidal or S-shaped curve*) at 50% plasmolysis (see Figure 19 on p. 23)

A line of best fit is added by eye. You should use a transparent plastic ruler to aid you (and remember to use a pencil to allow correction). When judging the position of the line, the following rules should be applied:

- There should be approximately the same number of points on each side of the line.
- The points should be evenly distributed, either side of the line, both at the top and the bottom of the line.
- The line should be near as many points as possible.

It is important to emphasise that it is not necessary to connect any points when you are constructing a best-fit line. Do *not* simply join the first and last points.

> **Practical tip**
>
> Not all lines of best fit go through the origin. You need to consider whether an independent variable with a value of 0 would result in a value of 0 for the dependent variable. If so, the line goes through the origin.

## Scaling

You do not need to start graphs at 0, 0. For example, if when measuring pulse rate before and after exercise all rates vary between 70 and 120, then you might well start the y-axis scale at 50 or 60, giving more space to plot the graph and making trends clearer — see Figure 35.

Sometimes two dependent variables are plotted on the same graph. If the scales for the dependent variables are quite distinct then two scales should be devised, preferably one on the left and one on the right — see Figure 36.

These aspects of scaling also apply to other graphical forms.

> **Exam tip**
>
> Always choose a scale with multiples of 2, 5 or 10. Avoid using scales that are difficult to work with, such as intervals of 3.

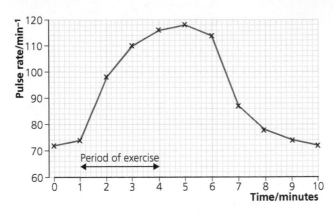

**Figure 35** The effect of exercise on pulse rate

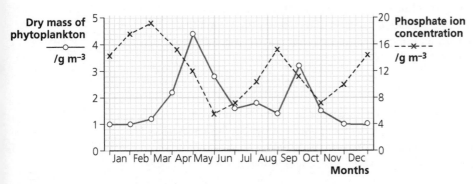

**Figure 36** Changes in phosphate ion concentration and phytoplankton biomass in a lake

## Identifying trends

When asked to interpret graphs, concentrate on the overall shapes of the patterns or trends. Do not worry about minor fluctuations. Some of the more common patterns are shown in Figure 37, along with the terms used to describe them.

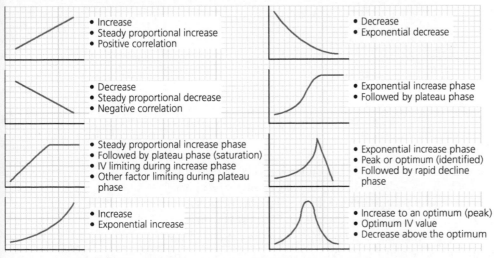

**Figure 37** Some common trends in biology, with descriptive terms

Keep your description of the trends and patterns concise. In describing a trend, do not restate every result. However, it can be helpful to quote one or two figures to help the description. For example, at what point does the curve level off or what is the optimum value? If you have several curves plotted on the same graph, try and pick out comparative points. What have the curves got in common and how do they differ? Remember that you are aiming to produce an overview.

### Calculating rates

Where time is the independent variable, you may be asked to calculate the rate where there is a linear relationship. The line is represented by the equation $y = mx + c$ (where $m$ = the gradient of the line and $c$ = the intercept on the $y$-axis), though the rate is simply determined by taking readings from the graph and using these to find the change in value and the time taken.

For example, the investigation of the effect of hydrogen peroxide concentration on the rate of catalase activity will give a linear relationship (with the enzyme in excess). Reading off the value of oxygen released (in $cm^3$) at two times on the graph and dividing the difference in oxygen production by the time interval (in seconds) calculates the rate of reaction ($cm^3 s^{-1}$).

## Histograms

A histogram is used to plot frequency distribution with continuous data. It would be used, for example, to show the frequency distribution of leaf width of wild garlic leaves. The leaf widths are grouped into classes. The frequency for each class is shown as a column, and adjacent columns should touch. An example is shown in Figure 38.

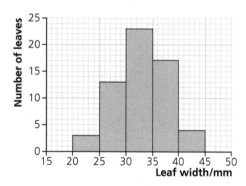

**Figure 38** The frequency distribution of leaf width in a population of wild garlic (*Allium ursinum*)

## Bar graphs (bar charts)

A bar graph is used when the independent variable is not numerical. The bars are of equal width and they do not touch. It would be used, for example, to show the effect of different colours of light on the rate of photosynthesis. In the experiment, shoots of the Canadian pond weed (*Elodea canadensis*) were exposed to light using various colour filters, and the time taken for the release of 20 bubbles from the cut end of the stem was recorded. The bar graph is shown in Figure 39.

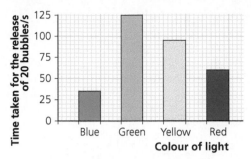

**Figure 39** The effect of different colours of light on the time taken for the release of 20 bubbles by the Canadian pond weed (*Elodea canadensis*)

# Scatter graphs (scattergrams)

A scatter graph or scattergram shows the relationship between two variables. The points on the graph are not joined up but left as simple crosses. It would be used, for example, to determine if there was a relationship between the size of heather plants and the pH of the soil. In the investigation, ten heather plants were chosen at random and their size estimated by measuring the diameter of the main stem, while the pH of the top centimetre of soil under the centre of each plant was also determined.

In this example (see Figure 40), soil pH may be influencing the growth of the heather, but it might just as well be that the heather is causing a more acidic soil pH. The axes could, indeed, be drawn the other way round. There is no obvious IV.

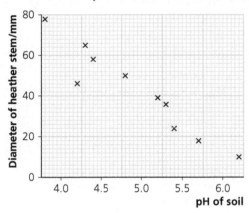

**Figure 40** The relationship between soil pH and diameter of heather (*Calluna vulgaris*) stems

The way in which the points fall on the graph shows the trend—see Figure 41.

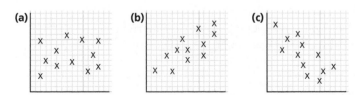

**Figure 41** Three possible trends on a scattergram: (a) no correlation; (b) positive correlation; and (c) negative correlation

> **Practical tip**
>
> Remember that correlation does not mean that changes in one variable *cause* changes in the other variable.

> **Practical tip**
>
> The stronger the correlation, the more the data points are clustered along a line.

## Other graphical forms

Two other graphical forms are to be found:

- **pie chart** — shows the data as portions of a whole
- **kite diagram** — shows the relative density of a species along a transect (see Figure 28 on p. 33)

# ■ AS Unit 3 assessing practical skills

AS Unit 3 includes a series of internally assessed practical tasks, and a 1-hour written examination assessing practical skills. Together these assessments are allocated 71 marks, constituting 25% of the AS award. Since AS represents 40% of A-level, this contributes 10% to the final A-level outcome.

Overall in the AS Unit 3 assessments, the approximate marks allocated for each assessment objective (AO) are:

AO1     Knowledge and understanding: 28

AO2     Application of knowledge and understanding: 30

AO3     Analysis, interpretation and evaluation of scientific information, ideas and evidence: 13

## Practical tasks internally assessed

You will be asked to complete at least seven practical tasks selected by your teacher (from a list provided on p. 38 of the specification). These tasks will test your ability to implement practical procedures — implementation is not a skill that can be assessed in a written paper.

You must make a record of the practical tasks that you have carried out. Your work will be assessed by your teacher and made available for presentation to CCEA for moderation. Each task will earn up to 3 marks so that in total 21 marks are available. This constitutes 7.4% of the AS award (and contributes 3% to the final A-level outcome).

## Practical skills written examination

The AS Unit 3 examination constitutes 17.6% of the AS award (and contributes 7% to the final A-level outcome). The paper lasts 1 hour and is worth 50 marks. It consists of between 7 and 10 structured questions, which will vary in length and style. Questions will assess your understanding of AS practical skills and your ability to apply them to familiar and unfamiliar contexts.

# ■Questions & Answers

This section consists of questions covering a range of different practical tasks. Questions such as these may appear in the exam papers for Unit 1 or Unit 2, though the Unit 3 written paper is designated for assessing 'Practical Skills'.

Following each question, there are answers provided by two students of differing ability. Student A consistently performs at grade A/B standard, allowing you to see what high-grade answers look like. Student B makes a lot of mistakes — ones that examiners often encounter — and grades vary between C/D and E/U.

Each question is followed by a brief analysis of what to look out for when answering the question (shown by the icon ⓔ). All student responses are then followed by comments (preceded by the icon ⓔ). They provide the correct answers and indicate where difficulties for the student occurred, including lack of detail, lack of clarity, misconceptions, irrelevance, poor reading of questions and mistaken meanings of examination terms.

## Biological molecules
### Question 1 Benedict's and biuret tests

**A student was given four test tubes containing the solutions shown in the table below. The student carried out Benedict's and biuret tests on samples of each solution. Complete the table to show the results of each test. Use '✓' for a positive result and 'X' for a negative result.**

(4 marks)

| Solution in test tube | Result of Benedict's test | Result of biuret test |
|---|---|---|
| Casein, a protein | | |
| Fructose | | |
| Sucrase and sucrose that had been incubated for 24 hours | | |
| Sucrose | | |

Total: 4 marks

ⓔ This should be straightforward if you have learned your tests for biological molecules.

| Student A | | | |
|---|---|---|---|
| **Solution in test tube** | **Result of Benedict's test** | **Result of biuret test** | |
| Casein, a protein | X | ✓ | ✓ |
| Fructose | ✓ | X | ✓ |
| Sucrase and sucrose that had been incubated for 24 hours | ✓ | ✓ | ✓ |
| Sucrose | X | X | ✓ ⓐ |

ⓔ **4/4 marks awarded** ⓐ 1 mark is given for each correct row.

**Student B**

| Solution in test tube | Result of Benedict's test | Result of biuret test | |
|---|---|---|---|
| Casein, a protein | ✗ | ✓ | ✓ |
| Fructose | ✗ | ✗ | ✗ |
| Sucrase and sucrose that had been incubated for 24 hours | ✓ | ✗ | ✗ |
| Sucrose | ✗ | ✗ | ✓ [a] |

ⓔ **2/4 marks awarded** [a] Student B has not realised that fructose is a reducing sugar; and while knowing that sucrase will act on sucrose (a non-reducing sugar) to release reducing sugars (fructose and glucose), has not considered that sucrase, as an enzyme, is a protein providing a positive result when tested with biuret.

## Question 2 Amino acid chromatography

The three peptides shown below were hydrolysed and the amino acid mixtures produced were spotted on the origin line of a strip of chromatography paper.

    X    lysine – alanine

    Y    alanine – glycine

    Z    glycine – isoleucine – lysine

The resultant chromatogram is shown below.

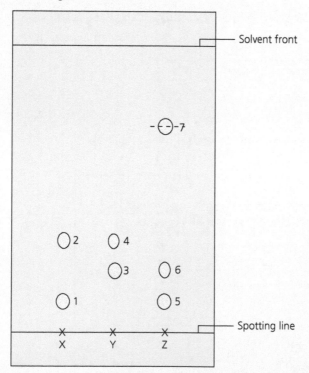

(a) Identify the amino acids listed 1 to 7. (4 marks)

(b) (i) Calculate the $R_f$ value for amino acid 7. The dotted line through this spot indicates its mid-point, which is used to measure the distance it has travelled. (2 marks)

(ii) Amino acid 7 and leucine have the same $R_f$ value in the solvent used for this chromatogram. Explain how they might be distinguished. (2 marks)

(c) Amino acid chromatography requires that a number of precautions be carried out. State two precautions required when:

(i) preparing the chromatography paper (2 marks)

(ii) spotting the amino acid mixtures (2 marks)

(iii) running the chromatogram (2 marks)

Total: 14 marks

ⓔ The question requires an understanding of a standard practical (amino acid chromatography) with an initial, relatively simple, problem-solving exercise in part (a). In (b) (i), remember to show your working as 1 mark may be available for a correct arithmetic operation, even if the final answer is wrong.

---

**Student A**

(a) 1 & 5 lysine; ✓ 2 & 4 alanine; ✓ 3 & 6 glycine; ✓ 7 isoleucine. ✓ ⓐ

(b) (i) $R_f$ = 54 mm/75 mm ✓ = 0.72 ✓

(ii) By using a different solvent. ✓ To be separated in the new solvent, the amino acids must have different solubilities in it so that they run at different rates. ✓ ⓐ

(c) (i) Mark the spotting line in pencil. ✓ Handle the chromatography paper only at the edges or wear disposable gloves. ✓

(ii) A fine pipette is used to spot the amino acids onto a marked point on the spotting line. ✓ The spot is dried and the process is repeated a number of times to produce a concentrated spot. ✓

(iii) The spotting line must not be allowed to dip into the solvent. ✓ The solvent front must not be allowed to reach the end of the paper. ✓ ⓐ

---

ⓔ **14/14 marks awarded** ⓐ All correct.

---

**Student B**

(a) 1 lysine; 2 alanine; 3 glycine; 4 alanine; ✓ 5 lysine; ✓ 6 glycine; ✓ 7 isoleucine. ✓ ⓐ

(b) (i) $R_f$ = 0.61 ✗ ⓑ

(ii) Run the amino acids in a different solvent. ✓ ⓒ

(c) (i) Use disposable gloves to handle the paper so chemicals from sweat do not contaminate it. ✓ Use a pencil to mark the points at which the amino acids will be spotted. ✓ d

(ii) The spot must be kept small so use a fine tube like a melting-point tube for the spotting process. ✓ After drying, further spots of the amino acids are added. ✗ e

(iii) While spraying with ninhydrin to show up the amino acids do so in a fume cupboard ✗ and wear protective gloves, mask and goggles. ✗ f

ⓔ 8/14 marks awarded a All correct for 4 marks. b It looks as though Student B used the distance of the solvent front from the bottom of the paper (rather than from the spotting line) as the denominator, though as no working has been shown no marks can be awarded. c No explanation has been provided for why using a different solvent might separate the amino acids, so only 1 mark awarded. d Both points correct. e The first point is correct, but the second is not — it is not extra spots which should be added, but further drops onto the original spot, with each dried after being added (to achieve a concentrated spot). f These are incorrect since they describe precautions for developing the chromatogram, not for running it. Always read questions carefully.

# Enzymes

## Question 3 The effect of enzyme concentration on the rate of reaction

The milk protein casein is broken down by protease enzymes such as trypsin. The white cloudy appearance of the milk is replaced by a clear solution. $2\,cm^3$ of different concentrations of trypsin was added to $8\,cm^3$ of milk suspension and the time taken for the cloudiness to disappear was recorded.

The rate of reaction was calculated as $\frac{1}{t} \times 100$, where $t$ = time taken for the cloudiness to disappear. The results are shown in the table below.

| Concentration of trypsin /% | Time taken for cloudiness to disappear ($t$)/s | Rate of reaction $\left(\frac{100}{t}\right)$ /s$^{-1}$ |
|:---:|:---:|:---:|
| 0.0 | 0 | 0.00 |
| 0.2 | 167 | 0.60 |
| 0.4 | 63 | 1.59 |
| 0.6 | 53 | 1.89 |
| 0.8 | 33 | 3.03 |
| 1.0 | 29 | |

(a) Calculate the rate of reaction for 1% trypsin. Show your working. (2 marks)

(b) Plot the results, using an appropriate graphical technique. (5 marks)

**(c)** Describe and explain the trend shown in the graph. (4 marks)

**(d)** Outline how you would control two variables which might otherwise affect the results. (2 marks)

**(e)** In a similar experiment, different concentrations of trypsin were made up by serial dilution of a 1% stock solution. Explain the term 'serial dilution' and describe how you would undertake this. (2 marks)

Total: 15 marks

 There are a number of skills being tested in this question: a calculation in (a); graphical construction in (b); describing and explaining a graphical trend in (c) — in describing the trend use the data in the table as well as the graph; controlling experimental variables in (d); and providing a definition, with exemplification, in (e).

---

**Student A**

**(a)** The rate of reaction = 100 ÷ 29 ✓ = 3.45 ✓ a

**(b)**

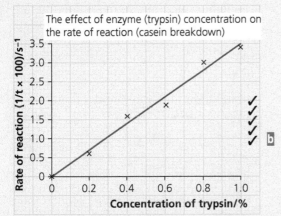

The effect of enzyme (trypsin) concentration on the rate of reaction (casein breakdown)

**(c)** The rate at which casein is broken down increases as the trypsin concentration increases. ✓ This is a linear increase with the rate of reaction approximately doubling as the enzyme concentration is doubled. ✓ As the concentration of enzyme increases, there are more active sites available, ✓ and there are more collisions between trypsin molecules and the casein molecules. ✓ More enzyme–substrate complexes can form. c

**(d)** pH would be controlled using a buffer solution at a set pH. ✓ Temperature would be controlled. ✗ d

**(e)** A serial dilution is a set of dilutions in a geometric series. ✓ In a ten-fold serial dilution, 1 cm³ 1% trypsin solution is added to 9 cm³ water to produce a 0.01% solution, 1 cm³ 0.01% solution is added to 9 cm³ water to produce 0.001% solution, and these steps repeated. ✓ e

(e) **14/15 marks awarded** [a] Correct. [b] Student A gets full marks: an appropriate caption is included; the axes are the right way round with the independent variable on the *x*-axis; the graph is well scaled with points accurately plotted; labels and units of measurement are shown; and a line of best fit is drawn (appropriate here since the graph includes six points, though without replication short, straight lines would also have been appropriate). [c] A full description and explanation of the relationship. [d] The first statement is correct, but Student A makes no reference to how temperature would be controlled — using a thermostatically controlled water bath. [e] Correct and well exemplified.

## Student B

**(a)** 3.45 ✓✓ [a]

**(b)**

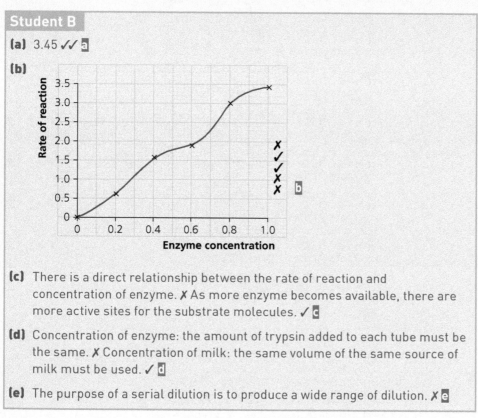

**(c)** There is a direct relationship between the rate of reaction and concentration of enzyme. ✗ As more enzyme becomes available, there are more active sites for the substrate molecules. ✓ [c]

**(d)** Concentration of enzyme: the amount of trypsin added to each tube must be the same. ✗ Concentration of milk: the same volume of the same source of milk must be used. ✓ [d]

**(e)** The purpose of a serial dilution is to produce a wide range of dilution. ✗ [e]

(e) 6/15 marks awarded [a] Student B has not followed the instruction to show working, but examiners will often give full marks even if working is not shown. [b] 2 marks are awarded for accurately plotted points in a well-scaled graph, and for putting the independent variable on the *x*-axis. However, there is no caption, there are no units of measurement and the line drawn is totally inappropriate. [c] The student has said that there is a 'direct relationship' but has not said what it is, and there is no reference to the data and the rate of increase. Further, while 1 mark is awarded for correctly referring to active sites, there is no further development of the explanation. [d] The first answer is the independent variable in this investigation and it is changed each time rather than being kept the same throughout. The second answer is correct for 1 mark. [e] This is the purpose but it does not answer the question — see Student A's answer.

# Question 4 Immobilised lactase in a reaction column

A solution of the enzyme lactase is mixed with a 2% sodium alginate solution.

**(a)** $500\,cm^3$ of sodium alginate solution was required. How many grams of sodium alginate powder would be added to $500\,cm^3$ of water to produce a 2% solution? (1 mark)

The mixture is drawn into a syringe and dropped, using constant pressure, into a solution of calcium chloride forming beads of alginate containing lactase. The faster the rate of dropping, the smaller will be the beads. The beads are placed in a glass column as shown in the apparatus below. Milk containing the disaccharide lactose (a reducing sugar) is trickled through the reaction column. The immobilised lactase hydrolyses the lactose in milk, forming the products glucose and galactose.

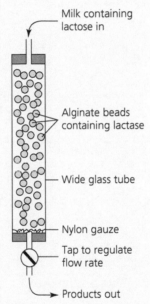

Milk containing
lactose in

Alginate beads
containing lactase

Wide glass tube

Nylon gauze

Tap to regulate
flow rate

Products out

**(b) (i)** Explain why Benedict's test cannot be used to determine that the reaction is taking place. (1 mark)

**(ii)** What test may be used to determine that lactose was hydrolysed? (1 mark)

**(c)** Suggest one advantage of immobilising the lactase used in the reaction column. (1 mark)

**(d)** Altering the size of the beads of immobilised lactase or the flow rate of milk through the reaction column affects enzyme activity and product formation.

**(i)** Using larger beads will decrease enzyme activity. Suggest why. (1 mark)

**(ii)** A slower flow rate will increase the amount of product formed. Suggest why. (1 mark)

Total: 6 marks

🅮 Part (a) is about testing for sugars and you will need to use the information provided, while the rest of the question examines your understanding of enzyme immobilisation.

---

**Student A**

**(a)** A 2% solution is 2 grams in 100 cm³, so 10 g need to be added to 500 cm³ water to make a 2% solution. ✓ a

**(b) (i)** Because lactose, the substrate, is also a reducing sugar. ✓

**(ii)** Clinistix tests specifically for the glucose produced. ✓ a

**(c)** The enzyme does not mix with the product so does not have to be removed. ✓ a

**(d) (i)** Larger beads present a smaller total surface area over which enzymes and substrate may interact, reducing enzyme activity. ✓

**(ii)** A slower rate will provide more time for the enzyme and substrate to interact increasing product formation. ✓ a

---

🅮 **6/6 marks awarded** a All correct.

---

**Student B**

**(a)** 20 g ✗ a

**(b) (i)** Because galactose is not a reducing sugar. ✗ b

**(ii)** A glucose-specific test-strip. ✓ c

**(c)** The enzyme can be used continuously in the reaction column. ✓ d

**(d) (i)** If the beads are large there will be fewer of them. ✗

**(ii)** Slower flow rate allows time for the substrate to collide and react with enzyme molecules in the beads. ✓ e

---

🅮 **3/6 marks awarded** a 20 g in 500 cm³ would produce a 4% solution. b Both galactose and glucose are reducing sugars, but the problem is that so is lactose, the substrate, information given in the question. c This is fine, since the specification does not provide a particular example. d Correct. e The question is not answered in (i), though correct in (ii).

# Cells

## Question 5 Plant cell ultrastructure

**The transmission electron micrograph (TEM) opposite shows parts of neighbouring plant cells in a young root.**

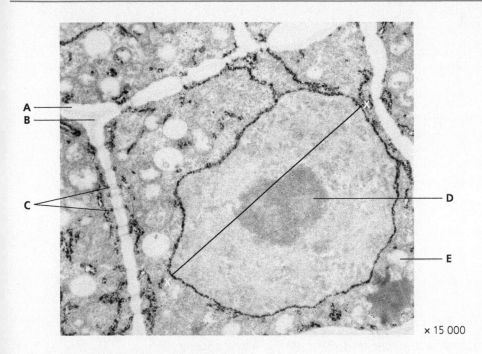

× 15 000

**(a)** Identify the structures labelled A to E. (5 marks)

**(b)** The magnification of the TEM is × 15 000. Calculate the true width, in μm, of the organelle along the line X–X. Show your working. (3 marks)

**(c)** Electron micrographs provide more information about the ultrastructure of cells because they have greater resolution than light microscopes. Explain the difference between resolution and magnification. (2 marks)

Total: 10 marks

ⓔ Part (a) is straightforward — you just need to take care to get the names right. The calculation of true size in part (b) should be a well-practised mathematical skill. Think carefully about the terms used in part (c) before starting to write — you must be precise about how the two terms are distinguished.

---

**Student A**

**(a)** A cellulose cell wall; ✓ B middle lamella; ✓ C plasmodesmata; ✓ D nucleolus; ✓ E mitochondrion ✗ ⓐ

**(b)** Measured length = 66 mm ✓ = 66 000 μm; ✓ 66 000 ÷ 15 000 = 4.4 μm ✓ ⓑ

**(c)** Resolution is the ability to distinguish two small objects — the smaller the objects you can distinguish, the greater the resolution. ✓ Magnification is how many times greater the image is than the object. ✓ You can keep on increasing magnification, but if the resolution stays the same you cannot see any more detail. ⓑ

---

ⓔ **9/10 marks awarded** ⓐ A to D are correct, but E is a vacuole, not a mitochondrion, since it is not covered in an envelope and there are no cristae apparent. ⓑ All correct.

Student B

**(a)** A cell wall; ✓ B lamella; ✗ C plasmodesmata; ✓ D nucleus; ✗ E vacuole ✓ a

**(b)** Measured length = 7 cm ✗ = 7000 μm; ✗ 7000 ÷ 15 000 = 0.47 μm ✓ b

**(c)** Resolution means how much detail you can see, and magnification means how much bigger something is. ✓ c

ⓔ 5/10 marks awarded a A, C and E are correct; in B, the answer is not sufficiently precise as lamella is the term used for an internal membrane in a chloroplast; D is within the nucleus but more precisely the nucleolus. b Student B has measured the length of the distance between the two Xs which lie outside the organelle. Furthermore, the measurement is in cm not mm, and so the conversion to μm is incorrect. However, 1 mark is gained for dividing by the magnification. c The student seems to understand the meaning of each term but needs to be more precise in order to gain both marks.

## Question 6 Drawing cells and calculating cell size

The false-coloured scanning electron micrograph (SEM) below shows the lower epidermis of a tobacco leaf (*Nicotiana tabacum*). Colour enhancement results in the inner surface of the guard cells appearing red.

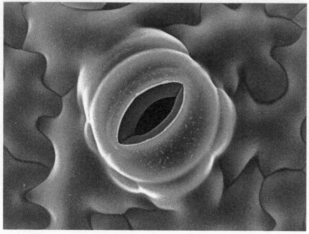

× 1000

**(a)** Make a labelled drawing, to the same scale, of the cells in the micrograph.        (4 marks)

**(b)** Explain the difference between a scanning electron micrograph and a transmission electron micrograph.        (2 marks)

**(c)** The leaf surface of a tobacco plant was also viewed under a light microscope fitted with an eyepiece graticule. At a magnification of × 500, the length of guard cells was measured at 6 eyepiece graticule units.

The eyepiece graticule was then calibrated, at the same magnification, using a stage micrometer in which each small division was 0.01 mm. The diagram opposite shows both scales.

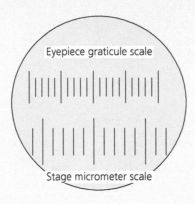

Eyepiece graticule scale

Stage micrometer scale

**Calculate the length of the guard cells in μm, using the above information.** (3 marks)

Total: 9 marks

ⓔ This question involves two skills that you should have practised: drawing cells in part (a); and, in part (c), calculating cell size using data from an eyepiece graticule and stage micrometer. In part (b) you need to be precise in distinguishing the two types of electron microscopy.

**Student A**

**(a)**

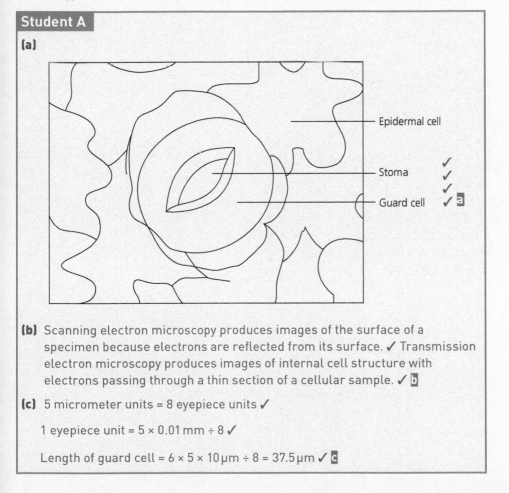

Epidermal cell

Stoma ✓
✓
✓
Guard cell ✓ a

**(b)** Scanning electron microscopy produces images of the surface of a specimen because electrons are reflected from its surface. ✓ Transmission electron microscopy produces images of internal cell structure with electrons passing through a thin section of a cellular sample. ✓ b

**(c)** 5 micrometer units = 8 eyepiece units ✓

1 eyepiece unit = 5 × 0.01 mm ÷ 8 ✓

Length of guard cell = 6 × 5 × 10 μm ÷ 8 = 37.5 μm ✓ c

**ⓔ 9/9 marks awarded** **ⓐ** This is a good biological drawing, with correct proportionality, continuous (non-sketchy) lines and appropriate labels. **ⓑ** Student A has presented a full distinction of SEM and TEM. **ⓒ** Correct and well laid out.

## Student B

**(a)**

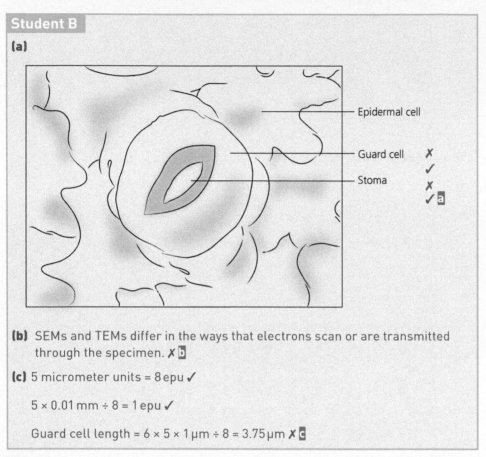

Epidermal cell

Guard cell ✗
✓
Stoma ✗
✓ⓐ

**(b)** SEMs and TEMs differ in the ways that electrons scan or are transmitted through the specimen. ✗ⓑ

**(c)** 5 micrometer units = 8 epu ✓

5 × 0.01 mm ÷ 8 = 1 epu ✓

Guard cell length = 6 × 5 × 1 μm ÷ 8 = 3.75 μm ✗ⓒ

**ⓔ 4/9 marks awarded** **ⓐ** Student B gets 2 marks for correct proportions and appropriate labels, but the drawing (including the guard cells) should not have shading while the lines are sketchy and often not continuous. **ⓑ** This vague answer reflects a lack of understanding. **ⓒ** The first two operations are correct for 2 marks, but Student B has not converted 0.01 mm to μm correctly.

# Question 7 Cell fractionation and organelle isolation

A sample of chloroplasts was required for an investigation. Some fresh spinach leaves were homogenised in ice-cold buffer solution. The mixture was strained through muslin and then spun in a test tube in an ultracentrifuge.

**(a) (i)** Explain why the spinach leaves were homogenised in a buffer solution.   (2 marks)

**(ii)** Explain why the buffer solution was:
- ice-cold
- isotonic   (2 marks)

**(b) (i)** The mixture was filtered before it was centrifuged. Suggest why.   (1 mark)

**(ii)** Explain how spinning the homogenised mixture in the ultracentrifuge would be used to obtain a fairly pure sample of chloroplasts.   (3 marks)

Total: 8 marks

ⓔ This question tests your understanding of a technique that you should be familiar with.

---

**Student A**

**(a) (i)** This keeps the pH constant, ✓ so that the enzymes do not denature. ✓

**(ii)** An ice-cold solution arrests enzyme activity. ✓ An isotonic solution stops organelles taking in water and bursting. ✓ ⓐ

**(b) (i)** This gets rid of cell debris. ✓

**(ii)** The first pellet will contain nuclei as these are the densest organelles. ✓ The supernatant will be put in a new tube and spun at a faster speed. ✓ The chloroplasts will be in the second pellet. ✓ They can be mixed with isotonic buffer and used in an investigation. ⓐ

---

ⓔ **8/8 marks awarded** ⓐ Full marks for well-expressed, correct answers.

---

**Student B**

**(a) (i)** To make sure the pH doesn't change. ✓ ⓐ

**(ii)** Being ice-cold stops bacteria growing. ✗ Being isotonic stops the cells bursting. ✗ ⓑ

**(b) (i)** To get rid of large particles. ✗ ⓒ

**(ii)** By centrifuging the mixture, so the heavy chloroplasts are in the pellet. ✗ ⓓ

---

ⓔ **1/8 marks awarded** ⓐ 1 mark for keeping the pH constant, but there is no reference to stopping the enzymes from denaturing. ⓑ The whole point of homogenising the tissue is to break the cells open — it is the organelles that need to be kept intact. ⓒ It needs to be clear that large particles will be parts of the tissue that have not been broken up in homogenisation. The term 'cell debris' is the best way to explain this. ⓓ Student B needs to explain that centrifugation separates organelles by *density*, and show understanding that chloroplasts will be in the second pellet.

# Membrane structure and function
## Question 8 Determining the average solute potential of plant epidermal cells

In an experiment to determine the average solute potential of onion epidermis, pieces of epidermis were immersed in sucrose solutions of different molarities. After 20 minutes, each piece of epidermis was mounted on a microscope slide using the sucrose solution in which it had been immersed, and viewed under low power.

**(a)** The sucrose solutions were made up from a $1\,mol\,dm^{-3}$ stock solution of sucrose. Calculate the volumes of distilled water and $1\,mol\,dm^{-3}$ sucrose solution required to produce $20\,cm^3$ of $0.3\,mol\,dm^{-3}$ sucrose solution. (2 marks)

The number of plasmolysed cells recognised within total number viewed (at least 50) was counted and the percentage of plasmolysed cells calculated. The results were plotted in a graph with % plasmolysis against molarity of sucrose solution. Alongside, a graph was plotted of the relationship between molarity of sucrose solution and solute potential.

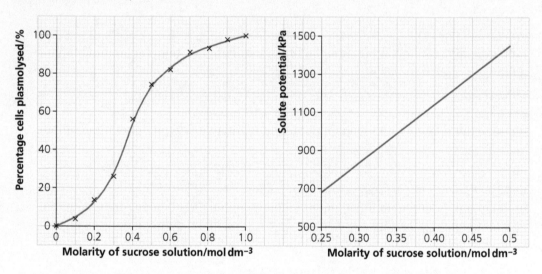

**(b)** In one sucrose solution used, of the onion epidermal cells viewed 17 were plasmolysed and 33 were not. Calculate the % plasmolysis for this result. (1 mark)

**(c)** Use the results graph to determine the molarity of sucrose solution equivalent to the average solute potential of the epidermal cells. Explain your reasoning. (2 marks)

**(d)** Use the calibration curve to determine the average solute potential of the epidermal cells in kilopascals. (1 mark)

Total: 6 marks

ⓔ The first parts involve calculations that you will have undertaken: in part (a) the production of a dilution; in (b) a percentage. In parts (c) and (d) you are asked to determine the average solute potential of epidermal tissue, a standard practical in your AS course.

---

**Student A**

**(a)** $6\,cm^3$ $1\,mol\,dm^{-3}$ sucrose solution ✓ added to $14\,cm^3$ water ✓ will produce $20\,cm^3$ $0.3\,mol\,dm^{-3}$ sucrose solution. ⓐ

**(b)** 17 plasmolysed out of a total of 50 cells viewed = 34%. ✓ ⓐ

**(c)** The average solute potential of the epidermal tissue is calculated at 50% plasmolysis and is $0.38\,mol\,dm^{-3}$ sucrose solution. ✓ At this point the average pressure potential is taken as zero and so the average solute potential of the tissue equals the solute potential of the external solution. ✓ ⓐ

**(d)** From the calibration, $0.38\,mol\,dm^{-3}$ equals $-1080\,kPa$. ✓ ⓐ

---

ⓔ **6/6 marks awarded** ⓐ Full marks from accurate and clear answers.

---

**Student B**

**(a)** To make a $20\,cm^3$ $0.3\,mol\,dm^{-3}$ sucrose solution you need to add $6\,cm^3$ $1\,mol\,dm^{-3}$ sucrose solution. ✓ ⓐ

**(b)** 17 out of 33 cells are plasmolysed = 51.5%. ✗ ⓑ

**(c)** At 50% plasmolysis on the graph, the average solute potential of the epidermal tissue equates to $0.38\,mol\,dm^{-3}$ sucrose solution. ✓ ⓒ

**(d)** $-1080\,kPa$. ✓ ⓓ

---

ⓔ **3/6 marks awarded** ⓐ Student B understands the calculation but needs to clearly state the volume of water to be used. ⓑ There are a total number of 50 cells viewed — 17 plasmolysed plus 33 not plasmolysed. ⓒ The average solute potential has been correctly determined but the student has failed to justify the reasoning behind this. ⓓ A correct use of the calibration curve.

## Question 9 The effect of alcohols on membrane permeability in beetroot

The purple pigments, betalains, in beetroot are confined in the cell vacuole. The pigments cannot pass through membranes, but can pass through the cellulose cell walls if the membranes are damaged. So disruption to membranes will cause pigments to leak out of the cells.

**(a)** Name the membranes which must be damaged for betalain pigments to leak out of the beetroot cells.

(2 marks)

An investigation is carried out to test the effect of three different alcohols — methanol, ethanol and propanol — on the integrity of the membranes within beetroot cells. Same-sized cores of beetroot are cut, placed in distilled water overnight to remove surface pigments, and then added to different concentrations of each of the three alcohols. The cores are left immersed for 30 minutes during which time pigments may pass into the alcohol solution. Then the cores are discarded. A colorimeter is used to measure the colour intensity of the pigments leaked into each alcohol solution. The results are shown in the graph below.

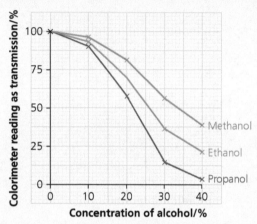

**(b)** Suggest an appropriate caption for the above graph. (2 marks)

**(c)** Describe the trends evident in the graph. (4 marks)

**(d)** Suggest an explanation for the effect of the alcohols on the integrity of the beetroot cell membranes. (2 marks)

**(e)** Suggest why there would be surface pigments on the cores of beetroot initially cut. (1 mark)

Total: 11 marks

ⓔ This question involves a practical which you will probably not have undertaken and so is designed to test the application of your understanding.

---

**Student A**

**(a)** The tonoplast. ✓ ⓐ

**(b)** The effect of different concentrations of three alcohols ✓ on the escape of pigments (measured using a colorimeter) from the sap vacuoles of beetroot tissue. ✓ ⓑ

**(c)** The colorimeter transmission readings decrease as more pigment is released from the beetroot tissue. ✓ All alcohols cause pigment to be released. ✓ Pigment release increases as the alcohol concentration increases. ✓ Propanol causes the greatest release and methanol the least. ✓ ⓑ

(d) Cell membranes are composed mainly of lipids and proteins. ✓ Alcohol will dissolve the lipids ✓ and denature the proteins. b

(e) When the tissue is cut some cells are disrupted and the cell contents are released. ✓ b

*e* 10/11 marks awarded a Student A has not realised that for the pigment to escape the cell-surface membrane must also be disrupted. b Well-phrased and correct answers.

---

**Student B**

(a) Plasma membrane ✓ and the membrane round the sap vacuole. ✓ a

(b) The effect of propanol, ethanol and methanol on the leaking of betalains from beetroot ✓ as measured using a colorimeter. b

(c) All alcohols cause transmission readings to fall. ✓ Transmission readings fall as betalain pigments leak out of the beetroot. ✓ Methanol causes the greatest disruption of the membranes and leakage of pigments. ✓ c

(d) The lipids which make up the bulk of the cell membranes ✓ are affected by the different alcohols. d

(e) This is the normal release of pigment from cells. ✗ e

---

*e* 7/11 marks awarded a Plasma membrane is an alternative name for the cell-surface membrane and while tonoplast is the scientific name for the membrane round the sap vacuole, it has been appropriately identified. b This is almost complete but Student B has ignored that different concentrations of the alcohols are being investigated. c There are 3 valid points, but the student has ignored the effect of the different concentrations. d 1 mark for implicating lipids, but it is not sufficient to simply say that there is an effect. e The pigments are only released when membranes are disrupted.

# The cell cycle

## Question 10 Mitosis and root tip squash

(a) (i) Outline the steps you would use to prepare a stained squash of cells from the root tip of a plant for examining stages of mitosis under a light microscope. (3 marks)

  (ii) A student prepared a stained root tip squash but failed to find any stages of mitosis. Suggest one reason why. (1 mark)

**(b)** The photomicrograph below shows cells from the tip of a plant root.

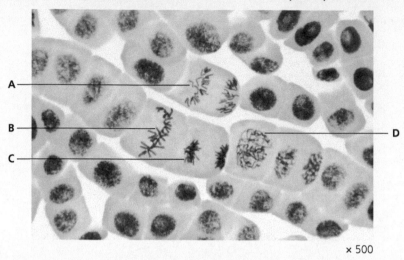

× 500

**(i)** Identify the stages of mitosis labelled A to D. (4 marks)

**(ii)** Put the letters A to D in the correct sequence, starting with the earliest stage. (1 mark)

Total: 9 marks

ⓔ Part (a) requires you to show an understanding of preparing a root tip squash to illustrate mitosis; in (i) notice that you are asked to 'outline' which means that only the main stages need be reported. In part (b), the identification of mitotic stages is required.

---

**Student A**

**(a) (i)** A root tip is heated at 60°C in a mixture of dilute hydrochloric acid, which separates the cells and makes squashing easier, ✓ and acetic orcein, which stains the chromosomes red. ✓ After 30 minutes, the tip is added to a microscope slide, a coverslip added, and the tissue squashed by pressing down on the coverslip. ✓

**(ii)** The student was looking at the wrong part of the root. ✓ ⓐ

**(b) (i)** A anaphase; ✓ B metaphase; ✓ C telophase; ✓ D prophase. ✓

**(ii)** D, B, A, C. ✓ ⓐ

---

ⓔ **9/9 marks awarded** ⓐ Full marks for accurate answers throughout.

---

**Student B**

**(a) (i)** A small section of root is added to a tube containing acetic orcein and heated in a water bath at 60° for 30 minutes. ✓ A short section is then placed on a microscope slide with some acetic orcein, a coverslip added and gently tapped with the blunt end of a pointed needle. ✓ ⓐ

**(ii)** The stain didn't work. ✗ ⓑ

**(b) (i)** A anaphase; ✓ B metaphase; ✓ C late anaphase; ✗ D interphase. ✗ **c**

**(ii)** D, B, A, C. ✓ **d**

**e** **5/9 marks awarded** **a** Student B has ignored the stage involving dilute HCl to soften the tissue and ease subsequent squashing. **b** Incorrect. **c** C is telophase since the chromosomes are no longer separating but clustered at each end. D is prophase since chromosomes are obvious even if long and stringy. **d** Correct.

# Gas exchange
## Question 11 Use of the respirometer

The diagram below shows a simple respirometer used to measure the rate of respiration of maggots.

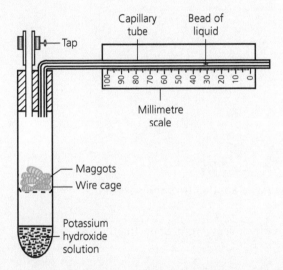

**(a)** Explain how, in the set-up shown, movement of the bead of liquid in the capillary tube is a measure of oxygen consumed by the maggots. (3 marks)

**(b)** The bead of liquid moved 70 mm in 5 minutes, the cross-sectional area of the capillary tube bore was 1 mm$^2$ and the maggots had a total mass of 0.5 g. Calculate the rate of oxygen consumption of the maggots. Give your answer in cm$^3$g$^{-1}$hour$^{-1}$. (3 marks)

Total: 6 marks

**e** In this question you need to show your understanding of how a simple respirometer works, and to undertake a calculation of oxygen consumption.

**Student A**

**(a)** As oxygen is consumed the pressure in the respirometer chamber decreases and the bead moves to the left. ✓ The pressure decrease is potentially compensated for by the production of carbon dioxide, ✓ though this is removed by the potassium hydroxide. ✓ **a**

**(b)** 70 mm × 1 mm$^2$ $O_2$ consumed by 0.5 g in 5 minutes

$\quad$ = 70 mm × 2 × 12 $O_2$ g$^{-1}$ h$^{-1}$ ✓

$\quad$ = 1680 mm$^3$ $O_2$ g$^{-1}$ h$^{-1}$ ✓ = 1.68 cm$^3$ $O_2$ g$^{-1}$ h$^{-1}$ ✓ a

ⓔ **6/6 marks awarded** a Full marks.

> **Student B**
>
> **(a)** Movement of the bead of fluid towards the tube with the maggots occurs because the pressure in the tube is reduced as oxygen is consumed. ✓ a
>
> **(b)** Oxygen consumption = 70 mm × 1 mm$^2$ × 2 per gram per 5 minutes ✓
>
> $\quad$ = 140 mm$^3$ × 12 g$^{-1}$ h$^{-1}$ ✓
>
> $\quad$ = 1680 ÷ 100 cm$^3$ g$^{-1}$ h$^{-1}$ = 16.8 cm$^3$ g$^{-1}$ h$^{-1}$ ✗ b

ⓔ 3/6 marks awarded a Student B has made no mention of carbon dioxide production and its absorption by the potassium hydroxide solution. b This was correct until the student made an incorrect conversion from mm$^3$ to cm$^3$, dividing by 100 rather than 1000.

# Plant structure

## Question 12 A tissue map of a xerophytic leaf

**The photomicrograph opposite shows a transverse section through part of a heather (*Erica*) leaf.**

**(a)** Draw a tissue map of the leaf section — a block drawing to show the tissue layers. $\qquad$ (4 marks)

**(b)** Identify three features which indicate that this is the leaf of a xerophytic plant and explain through annotation of the drawing how each feature acts as an adaptation. $\qquad$ (3 marks)

Total: 7 marks

ⓔ This question requires you to draw a tissue map (block diagram) — there are rules for achieving this which you will have practised. You should know the xerophytic features to identify.

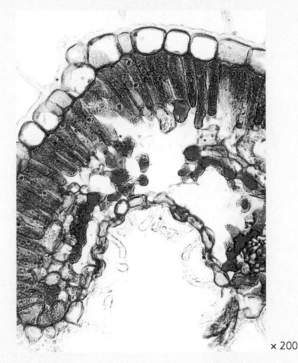

× 200

## Student A

**(a), (b)**

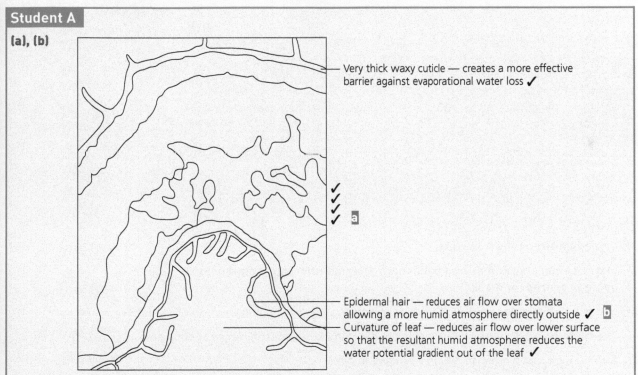

Very thick waxy cuticle — creates a more effective barrier against evaporational water loss ✓

Epidermal hair — reduces air flow over stomata allowing a more humid atmosphere directly outside ✓
Curvature of leaf — reduces air flow over lower surface so that the resultant humid atmosphere reduces the water potential gradient out of the leaf ✓

*e* **7/7 marks awarded** a An excellent drawing, representative of the micrograph, with all tissues shown, good proportionality and lines clear and continuous. b Annotations fully explanatory.

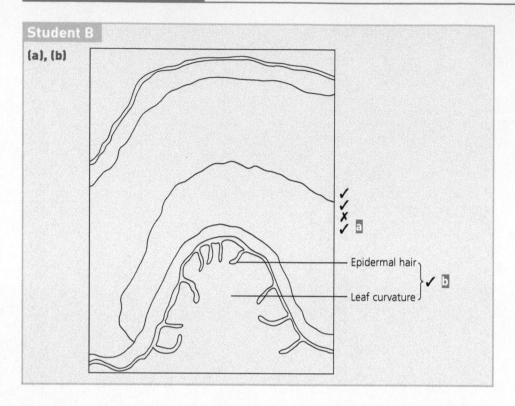

**Student B**

**(a), (b)**

Epidermal hair

Leaf curvature

**e** **4/7 marks awarded** **a** This is a good drawing representing the micrograph, showing all tissue layers (even if air spaces not shown) and with clear, non-sketchy lines. However, proportionality is not good enough, with the upper epidermis and its cuticle not that much bigger than the lower. **b** Student B has labelled two xerophytic features for 1 mark, though annotated explanations are not supplied.

# Circulatory system in mammals

## Question 13 Distinguishing aspects of the heart, blood vessels and blood cells

**Distinguish between the following.**

**(a)** The cut ends of the aorta and pulmonary arteries emerging from a heart being prepared for dissection.                                     (1 mark)

**(b)** The atrioventricular valves on the left and right sides of the heart.        (1 mark)

**(c)** The histological structure of an artery and a vein viewed in transverse section.   (1 mark)

**(d)** A monocyte and a lymphocyte in a blood smear.                   (1 mark)

Total: 4 marks

**e** This is a straightforward question requiring a comparison of different components of the circulatory system.

Student A

(a) The aorta has a much thicker wall and a slightly smaller lumen. ✓ a

(b) On the left side the AV valve has two flaps, while on the right there are three flaps. ✓ a

(c) The artery has a much thicker middle layer (tunica media) due to the large amounts of elastic fibres and smooth muscle. ✓ a

(d) A monocyte has a large kidney-shaped nucleus, while the lymphocyte nucleus is spheroid and larger, almost filling the content of the cell. ✓ a

e 4/4 marks awarded a Well-expressed and correct answers.

Student B

(a) The aorta has thicker walls ✓ to withstand the very high pressure of blood being pumped out to the body. a

(b) On the left side the atrioventricular valve is also called the mitral valve. ✗ b

(c) The artery has a thick middle layer containing an abundance of elastic fibres and smooth muscle. ✓ c

(d) The monocyte has a large nucleus which virtually fills the cell leaving little cytoplasm, while the lymphocyte has a kidney-shaped nucleus. ✗ d

e 2/4 marks awarded a The first part of the statement gets the mark. The rest is irrelevant. b Again, this is not an answer to the question, just an alternative name. c Correct. d Student B has got this the wrong way round, indicating the need to revise more thoroughly.

# Adaptations of organisms—ecology

## Question 14 Investigating plant distribution using a belt transect

A student sampled a hillside meadow, using an interrupted belt transect, for a number of plants: grasses (various species), meadow buttercup (*Ranunculus acris*) and rush (*Juncus effusus*). The percentage cover of the plants was determined and an estimate of soil moisture was also made at each quadrat position. The results are shown in the table overleaf.

| Distance along transect from top of hillside/m | Percentage cover of plants/% | | | Soil moisture index / arbitrary units |
|---|---|---|---|---|
| | Grass (various species) | Meadow buttercup (*R. acris*) | Rush (*J. effusus*) | |
| 0 | 40 | 30 | 0 | 2.0 |
| 10 | 45 | 50 | 0 | 2.0 |
| 20 | 60 | 30 | 0 | 2.5 |
| 30 | 50 | 40 | 5 | 3.0 |
| 40 | 60 | 30 | 10 | 3.5 |
| 50 | 40 | 30 | 20 | 4.0 |
| 60 | 40 | 25 | 35 | 4.5 |
| 70 | 40 | 15 | 40 | 5.0 |
| 80 | 30 | 15 | 55 | 5.5 |
| 90 | 10 | 10 | 80 | 5.5 |

(a) Why was an interrupted belt transect an appropriate technique for sampling the hillside meadow?

(2 marks)

(b) What do the results indicate about the habitat preference of the meadow buttercup (*R. acris*) and the rush (*J. effusus*)? Why is it not possible to make similar deductions about the habitat preferences of grasses?

(3 marks)

(c) Outline how the results would be presented using an appropriate graphical technique.

(3 marks)

Total: 8 marks

ⓔ This is a straightforward practical and one which you may have undertaken. Part (c) tests your understanding of graphical technique in displaying results.

> **Student A**
>
> (a) A belt transect is used because the investigation is of plant distribution along an environmental gradient. ✓ Interrupted sampling is used because the transect is long, approximately 100 metres. ✓ ⓐ
>
> (b) The meadow buttercup seems to prefer areas at the top of the slope where the soil moisture index is relatively low. ✓ The rush prefers the wetter conditions at the bottom of the slope, possibly because it has a competitive advantage there. ✓ Since grasses consist of many different species, no preference is obvious — some species may prefer drier conditions, while others prefer moister conditions. ✓ ⓐ
>
> (c) Distance along the transect is plotted on the *x*-axis as the independent variable. ✓ On the *y*-axis two scales are chosen — one for % cover of plant species and one for soil moisture index (say on the right-hand side). ✓ Points for the three plant groups and for soil moisture should be joined by short straight lines, since ecological data is highly variable (and there is no replication) so there is uncertainty about the accuracy of each point. ✓ ⓐ

ⓔ **8/8 marks awarded** ⓐ Very well-phrased and full answers throughout.

**Student B**

(a) Because the student is investigating the effect of an environmental gradient, soil moisture index down a slope, on plant distribution. ✓ a

(b) The meadow buttercup is more prevalent towards the top of the slope where the soil moisture index is below 3.0 to 4.0. ✓ The rush prefers the wetter conditions at the bottom of the slope. ✓ b

(c) Distance along the transect from the top of the hillside is the independent variable and placed along the *x*-axis. ✓ % plant cover is plotted on the *y*-axis. A separate graph with the same *x*-axis scale is plotted underneath with soil moisture index plotted on the *y*-axis. ✓ c

ⓔ 5/8 marks awarded a Student B has not explained that the belt transect should be interrupted because of the long length of the transect. b Again part of the answer is omitted — there is no suggestion about why it is not possible to make any deduction about the habitat preferences of grass species. c The first statement is correct and the student has presented an alternative way of displaying the soil moisture results along with the results of plant distribution. However, no comment has been made about joining up the points.

## Question 15 Simpson's index and habitat association

(a) In an investigation of the biodiversity of a stream, a section was randomly sampled to collect small aquatic animals. The results are shown in the table below, along with data for the calculation of Simpson's diversity index ($D$).

| Species of aquatic animal | Number of individuals ($n_i$) | $n_i(n_i-1)$ |
|---|---|---|
| Beetle larvae | 11 | 110 |
| Caddisfly larvae | 31 | 930 |
| Damselfly nymphs | 5 | 20 |
| Freshwater shrimps | 42 | 1722 |
| Lesser water-boatman | 13 | 156 |
| Mayfly nymphs | 103 | 10506 |
| Midge larvae | 20 | 380 |
| Stonefly nymphs | 23 | 506 |
| Water snails | 53 | 2756 |
| | $N = 301$ | $\Sigma n_i(n_i-1) = 17086$ |

(i) Calculate the value for Simpson's diversity index ($D$) for this section of the stream. Show your working. (2 marks)

The formula for calculating $D$ is:

$$D = \frac{\Sigma n_i(n_i - 1)}{N(N - 1)}$$

(ii) What can you deduce about the biodiversity of this section of the river? (2 marks)

(b) In a further investigation mayfly nymphs were sampled at 10 sites along the stream. The speed of the water current was also estimated at each sample site. The results are shown in the table overleaf.

| Number of mayfly larvae | 37 | 31 | 14 | 48 | 36 | 6 | 19 | 42 | 8 | 21 |
|---|---|---|---|---|---|---|---|---|---|---|
| Current speed/cm s$^{-1}$ | 72 | 66 | 31 | 94 | 79 | 19 | 59 | 91 | 12 | 42 |

   **(i)** What do the results indicate about the habitat preference of mayfly larvae? Suggest a reason for this preference.   (2 marks)

   **(ii)** Outline how the results would be presented using an appropriate graphical technique.   (3 marks)

Total: 9 marks

🄴 The question requires you to apply your understanding, first of Simpson's diversity index in part (a), and then of identifying a trend in part (b) where you will also need to demonstrate your understanding of graphical technique.

---

**Student A**

**(a) (i)** $D = \dfrac{\sum n_i(n_i - 1)}{N(N - 1)} = \dfrac{17086}{90300}$ ✓ $= 0.19$ ✓ 🄰

   **(ii)** $D$ is small, so there is a lot of biodiversity in the stream — zero represents infinite biodiversity. ✓ Also, the numbers are reasonably well spread among the 9 species sampled. ✓ 🄰

**(b) (i)** The results suggest that mayfly larvae prefer fast-flowing water. ✓ This may be because faster-flowing water contains more oxygen that the larvae need to keep active. ✓ 🄰

   **(ii)** A graph would be drawn with current speed plotted along the $x$-axis ✓ as the independent variable. ✓ Number of larvae are plotted along the $y$-axis. Since this graph could be regarded as a correlation there is no need to draw a line or lines between the points. ✓ 🄰

---

🄴 **9/9 marks awarded** 🄰 Full marks for comprehensive answers throughout.

---

**Student B**

**(a) (i)** $D = 17086 \div (300 \times 301)$ ✓ $= 0.189214$ ✗ 🄰

   **(ii)** The value for $D$ may vary between 0 and 1. Here the value is low. ✓ A low value of Simpson's index indicates that biodiversity is high. ✓ 🄱

**(b) (i)** Mayfly larvae are more numerous where the current speed is high. ✓ 🄲

   **(ii)** If it is possible to identify an independent variable, this must be placed on the $x$-axis. ✓ Since current speed cannot be dependent on the number of larvae and the converse seems to be true, then number of mayfly larvae is placed on the $x$-axis. ✓ 🄳

---

🄴 **6/9 marks awarded** 🄰 The initial calculation is correct, but the student has used too many decimal places when that level of accuracy is not appropriate. 🄱 The interpretation of $D$ is good. 🄲 Student B has identified the habitat preference but is unable to make any reasonable suggestion to explain it. 🄳 2 marks are awarded for explaining which variable goes on the $x$-axis, but no reasoning has been given about joining points or in this case not joining them.

# A2 practical skills

## ■ A2 Unit 1 practical work

### Microbiology

### Aseptic techniques

Aseptic techniques underpin all work in microbiology. These involve:

1 Working in a sterile bench-space
   - Work surfaces must be disinfected, before and after use.
   - Hands must be washed with antibacterial soap, before and after work.
   - Any transfer of microorganisms must be done close (say within 20 cm) to a Bunsen burner where warm air currents draw any airborne microorganisms upwards.

2 Accessing a glass container
   - On opening a flask, bottle or test tube containing agar or a microbial culture, its neck must be held over a Bunsen flame to kill any surface microbes, having removed the plug with the little finger curled towards the palm.

3 Accessing a Petri dish
   - An agar plate is prepared by pouring liquid sterile nutrient agar (sterilised by autoclaving) into a sterile Petri dish, having lifted its lid only at one side to allow access.
   - To add anything to the surface of the agar, such as a sample of the microorganism or an antimicrobial disc, the Petri dish lid must be opened only on one side for a minimal amount of time.

4 Flaming to sterilise instruments
   - A wire loop, sterilised by heating within a Bunsen flame, is used to transfer a microbial culture (from an existing culture on an agar slope or in a nutrient broth) to a nutrient agar plate, and sterilised afterwards.
   - Forceps, used in the transfer of antimicrobial discs, are also sterilised in a Bunsen flame (before and after use).
   - L-shaped glass spreaders are sterilised by dipping in ethanol and then flamed (before and after use).

5 Securing a Petri dish
   - The inoculated plate must be secured by fixing two or four short strips of adhesive tape at opposite ends of the dish.

Petri dishes, dropping (Pasteur) pipettes and spreaders are commercially available already sterilised (by irradiation).

**Aseptic** Free from contamination by microorganisms.

**Autoclave** Container using steam, produced under high pressure and temperature, to sterilise equipment — items can be kept dry by wrapping tightly in tinfoil.

Figure 1 shows the use of aseptic techniques to transfer bacteria from a broth culture to nutrient agar in a Petri dish.

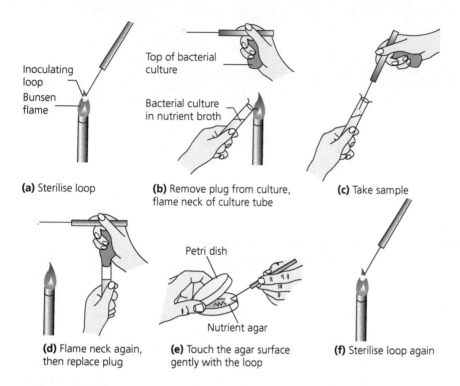

**(a)** Sterilise loop

**(b)** Remove plug from culture, flame neck of culture tube

**(c)** Take sample

**(d)** Flame neck again, then replace plug

**(e)** Touch the agar surface gently with the loop

**(f)** Sterilise loop again

**Figure 1** Using aseptic techniques to transfer bacteria from a broth culture to a Petri dish

## Effect of antibiotics on the growth of bacteria
### Investigating the effectiveness of different antibiotics

When a bacterial infection is diagnosed, it is important to determine which antibiotic will be most effective. In this investigation, you will be testing the effectiveness of several types of antibiotic on one species of bacterium.

A sterile Petri dish containing sterile nutrient agar is used. A sample is removed from a broth culture (only bacteria from the 'safe microorganisms' list provided by the Society for General Microbiology must be used) using a sterile dropper pipette, and a few drops are placed onto the surface of the agar in the Petri dish. A sterile spreader is then used to spread the inoculum over the entire surface of the agar. The inoculum is allowed to dry. Figure 2 shows the procedure for preparing a spread plate.

> **Practical tip**
> Bacterial cultures may be obtained in a liquid broth, or on an agar slope in a McCartney bottle.

**Culture** A microbial culture is a nutrient medium (broth or agar surface) in which a specific microorganism can reproduce and multiply.

**Inoculum** A sample amount of material containing microorganisms used to start a new culture.

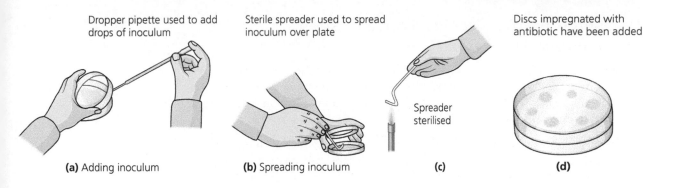

Figure 2 Preparing a spread plate (for the addition of antibiotic discs)

Sterile discs of filter paper (commercially available) are soaked in various antibiotics. Using sterile forceps, a disc is then placed onto the agar plate. For example, if comparing four different antibiotics, each would be placed centrally within separate quadrants of the dish. Alternatively, a Mast ring (a ring of paper with several 'arms', each treated with a different antibiotic) can be used. Label the underside of the plate with initials, date, name of microorganism and antibiotic used (or coded A, B, C etc.). The dish is taped and placed in an incubator at 20–25°C for 2–3 days.

After incubation, the plate will look opaque where bacteria have grown, but where the antibiotics have inhibited growth, clear zones called **inhibition zones** will be seen. The diameter of the inhibition zones may be measured in mm, or the area of these zones may be determined by placing mm-squared paper below the dish. This information may be used to decide which antibiotic is most effective at inhibiting the growth of the bacterium. Figure 3 shows the results for a particular Mast ring.

### Knowledge check 1

Identify the most effective antibiotic(s) from the results shown in Figure 3.

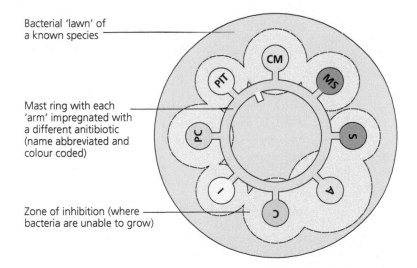

Figure 3 The inhibition zones for the different antibiotics on a Mast ring

### Exam tip

Be aware that an antibiotic found to be more effective against one species or strain of bacterium will not necessarily be the most effective against other strains or species.

## Investigating the effective concentration of an antibiotic

If a low concentration of an antibiotic inhibits growth then it can be regarded as effective in dealing with a bacterial infection. The lowest concentration of an antibiotic that inhibits the growth of a bacterial strain is called its **minimal inhibitory concentration** (**MIC**).

The precise concentration of antibiotic required to inhibit growth can be determined by preparing an inoculated agar plate (as in the previous investigation), adding an e-strip and incubating the culture. An e-strip has an exponential gradient of antibiotic concentrations on a paper strip. The result for a particular e-strip, exhibiting an elliptical zone of inhibition, is shown in Figure 4.

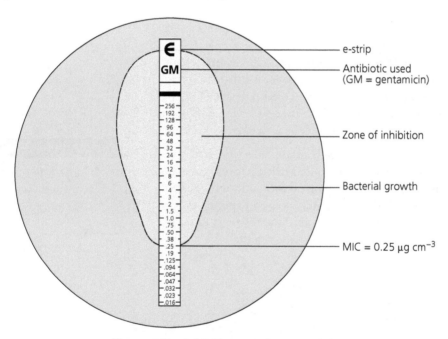

**Figure 4** The inhibition zone for an e-strip

Figure 4 shows the antibiotic's MIC for the inoculated bacterium. For patients with a specific infection, an antibiotic with a low MIC score will be administered. The dosage used might be several times the MIC score to ensure effective treatment and also prevent the evolution of antibiotic-resistant microbial strains.

## Preparing a streak plate to isolate single colonies

A sterile wire loop is used to remove a sample from the bacterial culture. The inoculum is smeared backwards and forwards across the surface of the agar (region A in Figure 5). Flame the loop, cool and, having turned the Petri dish 90°, streak the inoculum from A across the surface of the agar in parallel lines (B). This process is repeated twice more as shown in C and D. After a suitable period of incubation, colonies of bacteria will have grown on the agar.

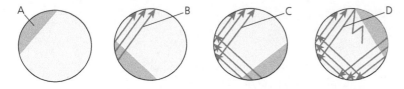

**Figure 5** Streak plating a bacterial inoculum to obtain single colonies

In the areas A, B and C they will probably form a continuous mat. In area D, however, you should find small colonies formed by the division of a single bacterial cell. Figure 6 shows the result of streak plating after inoculation of the agar plate.

**Figure 6** The results of streak plating

# Investigating the antimicrobial properties of plants

Some plants contain chemicals with antimicrobial properties, allowing them to resist infection from, for example, bacteria and fungi. Plants reputably with antibacterial or antifungal properties include those with culinary uses such as herbs and spices (e.g. garlic, thyme, cinnamon) or those producing essential oils (e.g. oil of cloves, lavender).

For investigating antibacterial activity, inoculated agar plates are produced, and sterile filter paper discs containing plant extracts are placed on the agar surface. To make a plant extract, 3 g of the plant material is crushed in $10\,cm^3$ ethanol. The paper discs are soaked in the extract and allowed to dry (within a sterile Petri dish) before using. The agar plates should be incubated for 2–3 days and measurements made of any inhibition zones.

If testing plant extracts for antifungal properties, a sterile agar plate is inoculated with a mould using a sterile wire loop, which is also used for spreading the mould across the surface. Fungi grow best at lower temperatures than bacteria, and so incubating at room temperature should suffice.

**Knowledge check 2**

Explain why control discs soaked in ethanol, and dried as for those containing the extract, should be used.

# Coordination in animals
## The eye
You need to examine photomicrographs of the mammalian eye and be able to identify: conjunctiva, cornea, iris, pupil, ciliary body, suspensory ligaments, aqueous and vitreous humours, retina (with rods and cones), choroid, sclera, blind spot and optic nerve. Often recognition of a feature will be based on its position within the eye.

## Skeletal, cardiac and smooth muscle
You need to examine photomicrographs of different muscle types: smooth, skeletal and cardiac. You should recognise each on the basis of characteristic features:
- smooth muscle — spindle-shaped single cells lacking striations
- skeletal muscle — multinucleate cells (since many cells are fused together) with obvious striations
- cardiac muscle — striated and branched forming a network, though individual cells are separated by intercalated discs

Recognition is aided by knowing the location of each type: smooth in the wall of the gut, bladder and blood vessels, and in the iris and ciliary muscles of the eye; skeletal attached to bones; and cardiac within the heart only.

You must also examine photomicrographs and electron micrographs of skeletal muscle to identify: myofibril, thick filaments of myosin, myosin heads, thin filaments of actin, sarcomere, A-band, I-band, H-zone, Z-line and M-line.

**Knowledge check 3**

Why is 'striated' not an accurate alternative name for skeletal muscle?

# Populations
## Investigating the growth of a yeast population using a haemocytometer
The population of a yeast culture can be investigated by adding a sample of yeast suspension to a flask of nutrient medium which has been previously sterilised. A suitable nutrient medium for yeast is 2% glucose solution (as carbon source) and mineral salts (e.g. ammonium as a source of nitrogen for amino acid synthesis, phosphate for nucleotide synthesis).

The flask containing the yeast cells and the nutrient medium is plugged and incubated at 25°C over several days. Samples are taken at intervals (at least twice a day) and the yeast cells counted using a **haemocytometer**. Cell density (cells mm$^{-3}$) is plotted against time (hours).

### Using a haemocytometer
A haemocytometer consists of a special glass slide with an accurately ruled, etched grid of precise dimensions (see Figure 7). The small, type-C square is $0.05\,\text{mm} \times 0.05\,\text{mm} = 0.0025\,\text{mm}^2$ in area. When the coverslip is correctly positioned over the counting grid, the depth of the counting chamber is 0.1 mm, and the volume of a type-C square is therefore $0.00025\,\text{mm}^3$ (or $1/4000\,\text{mm}^3$).

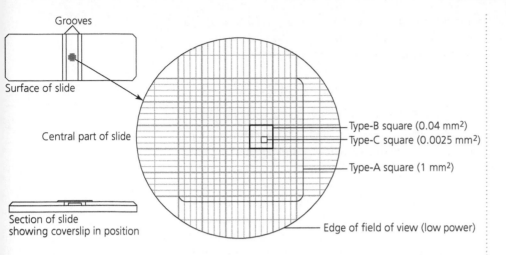

**Figure 7** A haemocytometer

To obtain a cell count, a representative sample of the yeast culture is placed beside the coverslip so that capillarity draws it under. The number of cells in 40 type-C squares is counted and totalled. Different numbers of squares may be counted, but 40 facilitates the calculation of cell density using the formula:

$$\text{number of cells per mm}^3 = \frac{\text{total number of cells in 40 type-C squares}}{40} \times 4000$$

The graph should show a phase of exponential increase followed by saturation (when the curve plateaus) and may even include a decline phase (depending on the duration of the experiment). An *explanation* of the exponential phase should indicate that resources are plentiful and not limiting yeast cell division. A plateau phase occurs when some factor is becoming limiting — such as lack of resources or accumulation of waste (ethanol or carbon dioxide which would cause the pH to fall). A decline phase occurs when many cells are dying (due to lack of resources or accumulation of waste) and not being replaced.

There are many factors to be careful about when doing this experiment:

- A representative sample of the yeast culture is obtained when the culture is shaken to ensure a homogeneous suspension before sampling.
- The sample on the haemocytometer must not enter the grooves, otherwise the procedure has to be repeated.
- When counting in the type-C squares, include those cells touching the top and left boundaries (sometimes called the north-west rule) and exclude those touching the right and bottom. This ensures that no cell can be counted twice.
- If there are very few cells in the sample (the yeast population is in an early stage of growth), type-A or type-B squares should be used as many type-C squares will be empty.
- If samples are taken when the yeast population has become dense, there may be too many cells to count. In this situation, an accurate dilution, say 1 in 10, should be made and the final count should then be multiplied by the dilution factor.

## Knowledge check 4

The average number of cells in a type-C square was determined as 8.15. Calculate the number of cells per $mm^3$.

## Knowledge check 5

Why is it a problem if the yeast sample enters the grooves on either side of the central platform?

● The haemocytometer gives a total count including both living (or viable) and dead (non-viable) cells. Living and dead cells may be distinguished by staining with methylene blue: dead cells are blue, while living cells are colourless.

## Estimating the size of an animal population using a simple capture–recapture technique

Many animals move and often, during daytime, remain hidden. The technique for estimating their population size is called **capture–mark–recapture** (also called **mark–release–recapture**). A sample of individuals is caught, counted and marked in some way — this is the first sample ($s_1$). These marked individuals are released back into their original location. After being allowed to mix with the unmarked individuals in the population, a second sample ($s_2$) is caught and counted and the number of marked individuals noted — these are the recaptures ($r$) since they were also caught on the first occasion. An estimate of the total population size (the **Lincoln index** or **Peterson estimate**) can then be calculated:

$$\text{population size} = \frac{s_1 \times s_2}{r}$$

The method of capture and marking needs to be appropriate for the animal species being investigated. For example, ground beetles can be caught in pitfall traps (see page 34) and marked by placing a minute drop of correction fluid or quick-drying waterproof paint on one of their hardened front wings or on the ventral surface, out of sight to most predators.

Using this technique for estimating population size relies on a number of *assumptions*:

● The mark should not harm the animal (this can be tested by keeping a sample of marked individuals in the lab to check for toxicity), it should persist over the sampling period, and it should not make the animal more obvious to predators or influence its behaviour. If these requirements are met, the probability of capturing a marked individual should be the same as that of capturing any member of the population.

● The marked individuals are completely mixed in the population, and enough time has elapsed between visits to the study area to allow this to happen.

● There are no significant gains or losses through immigration and emigration. This may be avoided by sampling a population isolated from other populations and ensuring that the two visits to the study area are close enough in time.

● There are no significant gains or losses through births and deaths. If the above requirements are met, then the proportion of marked individuals in the second sample is equivalent to the proportion of marked individuals in the whole population.

● If the estimate is to be reliable, a large sample should be marked initially — between 10 and 30% of the population.

### Knowledge check 6

A sample of 50 woodlice was removed from a study site, marked and released back into the population. Three days later, a second sample of woodlice was collected: 48 were unmarked and 12 marked. Calculate the estimate of the population size.

# ■A2 Unit 2 practical work

## Respiration
### Measuring the respiratory quotient using a respirometer

The **respiratory quotient** (**RQ**) is a measure of the ratio of carbon dioxide produced by an organism to the oxygen consumed over a given time period:

$$RQ = \frac{\text{volume of } CO_2 \text{ produced}}{\text{volume of } O_2 \text{ consumed}}$$

The volume of oxygen consumed by living organisms, such as bean seeds, is determined using a respirometer with potassium hydroxide to absorb the $CO_2$ produced (see pages 27–28). The volume of carbon dioxide produced is determined, for the same bean seeds, using a respirometer with water replacing the potassium hydroxide. This measures the net difference between $CO_2$ production and $O_2$ consumption and so, having already determined the volume of oxygen consumed, the volume of carbon dioxide produced can be calculated.

RQ values provide clues to the substrate being respired and the extent to which the tissue is respiring anaerobically (see the A2 Unit 2 Student Guide in this series, p. 14).

**Knowledge check 7**

Humans have an RQ of 0.85. Discuss.

## Using redox indicators to demonstrate dehydrogenase activity in respiration

During respiration, particularly during the Krebs cycle, dehydrogenase enzymes remove hydrogen which is taken up by hydrogen acceptors (NAD generally, but also FAD) which subsequently become reduced. Experiments can be devised in which the hydrogen released by dehydrogenase activity in living yeast cells is taken up by artificial hydrogen acceptors which change colour when reduced. Since the 'artificial' acceptors are one colour when oxidised (the normal state in an atmosphere containing oxygen) and another colour when reduced, they are called **redox indicators**. Three redox indicators are shown in Table 1.

**Table 1** Some redox indicators and their colours when oxidised and reduced

| Redox indicator | Colour when oxidised | Colour when reduced |
|---|---|---|
| Methylene blue | Blue | Colourless |
| Triphenyl tetrazolium chloride (TTC) | Colourless | Pink |
| Dichlorophenol indophenol (DCPIP) | Blue | Colourless |

Dehydrogenase activity in respiration can be investigated using a range of living material, such as seeds or yeast. A redox indicator is added to a suspension of the living material and observations are made of the colour change as the indicator is reduced. For example, using methylene blue, a measure can be made of how long it takes for the blue colour to disappear. A colorimeter could be used as a quantitative means of measuring the reduction of the indicator — percentage transmission should increase as the solution loses its blueness.

Redox indicators may be used in various experiments on dehydrogenase activity:

- a demonstration of hydrogen release in a living yeast suspension compared to a control containing 'boiled' yeast
- the effect on dehydrogenase activity of adding an intermediate of the Krebs cycle, such as succinate (the substrate for succinate dehydrogenase)
- the effect of malonate (a competitive inhibitor of succinate dehydrogenase) on dehydrogenase activity

# Photosynthesis
## Paper chromatography of plant pigments

Photosynthetic pigments can be separated and identified by paper chromatography (see pages 6–7). The pigments are extracted from a leaf and a concentrated spot is created on the chromatography paper. The different pigments are identified by their different colours and positions (see Figure 8). $R_f$ values are calculated by adapting the formula on page 7.

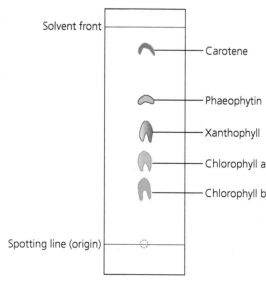

**Figure 8** Plant pigments separated on a paper chromatogram

The role of different pigments in absorbing light of different wavelengths may be discussed. It is also possible to analyse the pigments in different seaweeds.

## Using redox indicators to demonstrate the role of hydrogen acceptors in photosynthesis

During the light-dependent stage of photosynthesis, electrons are energised to reduce NADP. Using DCPIP to accept electrons demonstrates the activity of the light-dependent stage experimentally. As the DCPIP is reduced it turns from blue to colourless.

To demonstrate the activity of the light-dependent stage of photosynthesis, DCPIP is added to a suspension of chloroplasts, and the time taken to decolourise the DCPIP is measured. The loss of colour in the DCPIP is due to electrons produced by

**Knowledge check 8**

During an experiment on the respiration of yeast, using methylene blue as a redox indicator, students were directed to ignore the thin film of blue colour at the surface. Suggest why.

**Knowledge check 9**

Explain the role of different pigments in photosynthesis.

**Knowledge check 10**

DCPIP added to a suspension of whole plant cells turned from blue to colourless. Explain why this does not demonstrate solely the activity of the light-dependent stage of photosynthesis.

light-dependent reactions in the isolated chloroplasts. A control in which DCPIP is added to water would remain blue.

# DNA (gene) technology
## Extraction of DNA

DNA can be extracted from a range of sources. The following procedure uses an onion bulb.

1 **Releasing the cell contents.** Finely chopped onion, to which a little salt (NaCl) and detergent have been added, is homogenised using a mortar and pestle (or blender). Homogenisation disrupts the cell walls and the detergent causes cell membranes (cell-surface and nuclear membranes) to rupture, so that the cell contents are released in solution. The extract is filtered, giving a filtrate containing soluble proteins (sugars etc.) and DNA.

2 **Enzymic breakdown of proteins.** The filtrate is incubated with protease to break down the protein framework of the chromosomes and free the DNA.

3 **Precipitation of the DNA.** Ice-cold ethanol (or isopropanol) is added down the side of the beaker so that it floats on top of the filtrate. In the presence of the sodium and chloride ions (decreasing the solubility of DNA in water), ethanol causes the DNA molecules to coalesce and precipitate out of solution. Precipitated DNA fibres form at the water–ethanol interface, and can be drawn out on a rotated, fine glass rod.

4 **Washing and resuspending the DNA.** The precipitated DNA is 'washed' with a 70% ethanol solution to remove salts and other water-soluble impurities. The clean DNA is then resuspended in a buffer to ensure stability.

## Gel electrophoresis of DNA

An agarose gel tray is prepared with wells in the gel at one end. This is added to the electrophoresis apparatus with the well end placed towards the cathode (negative) end. The gel is just covered with buffer. A sample of the extracted DNA is placed in a well. The apparatus is connected to an electrical supply (80–150 V) and in the subsequent electrical field, DNA molecules, being negatively charged (due to the presence of phosphate), migrate towards the anode (positive) end. The resultant migrant DNA can be detected by staining with methylene blue.

If the DNA is treated with a restriction endonuclease enzyme, the fragments produced will be separated by gel electrophoresis, since the smaller fragments will migrate further in the electrical field — see Figure 9.

**Knowledge check 11**

Explain why detergent breaks down membranes.

**Exam tip**

Notice that this procedure differs from cell fractionation in which the aim is to keep the organelles intact. In this case it is essential to disrupt the organelles so that DNA is released from the nuclei.

**Knowledge check 12**

What is the role of salt in DNA extraction?

**Knowledge check 13**

The anode (positive pole) attracts molecules with what charge?

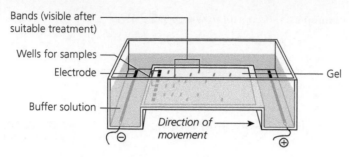

**Figure 9** Separating DNA using gel electrophoresis

# Kingdom Plantae

You need to be able to recognise key features of different plant groups using a range of sources: living and preserved specimens, prepared slides and photographs. The plant groups for study include:

- mosses
- ferns
- flowering plants

As a practical task, where the emphasis would be on demonstrating your manipulative skills, you could undertake a leaf scrape. This involves scraping the other layers of the leaf by repeatedly drawing a razor blade over a leaf held on a microscope slide until only the transparent epidermal layer and cuticle are left.

# Kingdom Animalia

You need to be able to recognise aspects of the body form of different animal groups using a range of sources: living and preserved specimens, prepared slides and photographs. The animal phyla for study include:

- Cnidaria (e.g. *Hydra* and jellyfish)
- Platyhelminthes (e.g. planarian and liver fluke)
- Annelida (e.g. earthworm and lugworm)
- Arthropoda (e.g. insect and spider)
- Chordata (e.g. mammal and bird)

**Exam tip**

Many features of the different animal or plant types may be studied through internet searches. Be careful to restrict your study to those features specified in the course.

As a practical task, where the emphasis would be on demonstrating your manipulative skills, you could undertake a dissection:

- insect mouthparts — see p. 115
- alimentary canal (digestive system) of a rat
- thorax of a rat

**Knowledge check 14**

The members of which plant division lack vascular tissue?

**Knowledge check 15**

An animal was found to be radially symmetrical. To which phylum would it belong?

# Dealing with data: logarithms and statistics

## Logarithms

Occasionally in investigations you will have a mixture of very small and very large numbers. This is the case with concentrations involving serial dilution (see pages 8–9) and in studies of population growth. In such circumstances, it is difficult to work out a satisfactory graphical scale which allows the small numbers to be accurately plotted. This is when a logarithmic, or log, scale may be used, where the intervals on the axis increase by an order of magnitude.

Logarithmic scales are especially useful for plotting data on the growth of microorganisms, since under ideal conditions the number of individuals increases exponentially. So a yeast population increasing, over equal time intervals, through 10, 100, 1000, 10000 to 100000 is more readily plotted on a log scale as 1, 2, 3, 4 and 5. This is not only easy to plot but turns an exponential curve (of ever increasing gradient) into a straight line.

## Statistical techniques

The use of statistical techniques is an important part of biology. There are two associated reasons for this:

- The data obtained is only ever really a sample of what could be obtained — the **population**. It is important that, as far as possible, the sample represents the population and that data collection avoids bias — it is important that the sample is random.
- Biological material is highly variable as a result of both genetic and environmental factors. The result is that you can reasonably expect two (or more) samples to differ, even if randomly selected.

Statistics is the use of mathematical methods to describe data and to determine the **probability** of events, such as whether differences are due to random factors (chance). Statistical techniques ensure that decisions are made in an **objective** way (as opposed to being subjective).

The statistical technique that is appropriate will depend on what you are dealing with:

- **Measured data** — where you have obtained results by measuring something (say, rate of reaction), repeated the results and calculated a **sample mean**. Note that even if you repeated results in a laboratory a number of times (e.g. five repeats for each pH in the investigation of amylase activity), the replicates still represent a sample of what you could potentially have done, say hundreds of repeats; and the calculated mean represents a sample mean.
- **Categorical data** — where you have obtained frequencies (counts) that fall within distinct categories, say the number of germinated seeds out of a total number used, or the number of different fruit fly variants in a genetic cross.

**Exam tip**

You may have seen a logarithmic scale in the A2 Unit 1 Student Guide in this series when answering question 5 (p. 76) — the concentration of auxin on the x-axis is shown on a logarithmic scale.

**Exam tip**

The logarithm of a number is the power to which the base must be raised to give that number. So the log to base 10 of 1000, which equals $10^3$, is 3. Check by entering 1000 into your calculator and pressing 'log'.

**Population** In statistics a population refers to a theoretical total of all the measurements or counts that might be made, e.g. an infinite number of measures of catalase activity, or all the limpets on a rocky shore.

**Probability** A measure of the likelihood that some event will happen.

**Objective** Based on observable facts that can be verified and not on personal judgement.

# Statistical analysis of sample means

The number of measurements that can be made is limited (for example, by time constraints or availability of equipment) and so the measurements represent a sample. All the measurements that might be made represent the population (whether real, e.g. a population of wild garlic plants, or imaginary, such as all the measurements of amylase activity that could potentially be taken). The sample provides an estimate for the population. There are two potentially important properties that summarise the sample of data collected:

- The sample mean (symbol $\bar{x}$) — an arithmetic measure of central tendency or average.
- The **standard deviation** — a measure of the variability (or spread or dispersion) of the data. As an estimate of the variability of the population, this is denoted by the symbol $\hat{\sigma}$.

A sample mean ($\bar{x}$) provides an estimate of the mean of the population from which the sample has been drawn. The population mean is referred to as the true mean (and denoted by $\mu$). How reliable is the sample mean, that is, how close does it lie to the true mean? Reliability depends on two things:

- The **sample size**, $n$. The bigger the sample size — the greater the replication — the more reliable will be the calculated mean.
- The variability of the data, as measured by $\hat{\sigma}$ (the estimate of the standard deviation). Greater control of variables will result in a more reliable mean.

Since sample means vary, it is important to consider how good an estimate any one sample mean might be. A sample mean will be more likely to lie close to the true mean if the sample size is large and the standard deviation is small. This is measured in a statistic called the **standard deviation of the mean** (also called **standard error of the mean**) with the symbol $\hat{\sigma}_{\bar{x}}$:

$$\hat{\sigma}_{\bar{x}} = \sqrt{\frac{\hat{\sigma}^2}{n}}$$

The standard deviation of the mean ($\hat{\sigma}_{\bar{x}}$) is a measure of how much the sample means would on average differ from the population mean. A small $\hat{\sigma}_{\bar{x}}$ value, relative to the magnitude of the sample mean, indicates a reliable sample mean — that is, one that lies close to the population mean.

# Statistical testing

There are three methods available for statistically comparing data:

- comparing sample means and their 95% confidence limits
- the $t$ test for comparing two sample means
- the $\chi^2$ test for comparing observed frequencies with those expected on the basis of theory

You need to be aware which analytical technique to use. Figure 10 should help you with this.

**Exam tip**

In laboratory experiments, where variables may be well controlled, then a minimum of three replicates might suffice, though five would be better. In ecological investigations, where variables cannot be readily controlled, then a replication of 30 may be required.

**Practical tip**

The statistics sheets used in A-level biology are provided at the back of the specification. You should download these. They include the equations that you will need to use (as well as $t$ and $\chi^2$ tables).

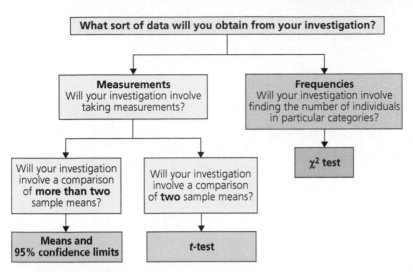

**Figure 10** Selecting the appropriate method of statistical analysis

## 95% confidence limits

Based on the sample data, it is possible to estimate the boundaries within which the true mean might lie. Since it is not possible to have absolute certainty, a convention of 95% probability has generally been accepted in biology. **95% confidence limits** are provided by:

$$\bar{x} \pm t(\hat{\sigma}_{\bar{x}})$$

where $t$ is determined from a table of $t$ values at $p = 0.05$ and $n - 1$ degrees of freedom (df).

95% confidence limits are particularly useful when plotted on graphs — the upper limit as a bar above the mean, the lower limit as a bar beneath. Since they determine, with a 95% probability, where the true mean lies, 95% confidence limits are used to compare different sets of data. If the 95% confidence limits of two samples overlap, this indicates that there is no significant difference between the means; if they do not overlap, this strongly *suggests* that there is a significant difference — the difference between the means is real and not due to chance. This decision can only truly be made by undertaking a $t$-test. However, it is the only way in which you can make decisions about significant differences when you are comparing more than two sample means.

**Worked example**

An investigation was carried out on the mass of brown trout (*Salmo trutta*) in each of three lakes (designated A, B and C).

For the data from lake A: the sample size is 30; the mean mass is calculated as 411 g; and the standard deviation (error) of the mean is calculated as 44.01. 95% confidence limits are calculated by the following steps:

1 As the sample size is 30, the degrees of freedom (df) is 29.
2 The tabulated $t$ value, at $p = 0.05$ and 29 df, = 2.045.
3 95% confidence limits (either side of the sample mean) = 2.045 × 44.01 = 90.
4 It is concluded, with a confidence of 95% probability, that the true mean of the population lies between 321 and 501 g (411 g ± 90).

➡

Comparable calculations for the mass of trout in lakes B and C were also made. Figure 11 compares the means and 95% confidence limits for the mass of trout from the three lakes.

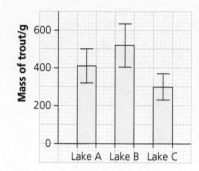

**Figure 11** A comparison of the mass of trout sampled from three lakes: means and 95% confidence limits plotted

You can see that, in comparing the mass of trout in lakes A and B, the limits *overlap*. So the samples are not significantly different — the difference between the sample means is due to chance. However, comparing the mass of trout from lakes B and C, you can see that the limits *do not overlap*. This suggests that the samples are significantly different — you need to carry out a *t*-test to confirm this.

**Knowledge check 17**

Regarding Figure 11, what might you suggest about the mean mass of trout in lakes A and C?

## *t*-test

The ***t*-test** (sometimes called **Student's *t*-test**) is a strong statistical procedure for comparing two sample means. Remember that sample means are expected to differ. The *t*-test allows you to determine if the difference is significant, that is, not just due to random factors or chance. The formula for calculating *t*, in terms of $\hat{\sigma}_{\bar{x}}$, is given as:

$$t = \frac{\bar{x}_1 - \bar{x}_2}{\sqrt{\hat{\sigma}_{\bar{x}_1}^2 + \hat{\sigma}_{\bar{x}_2}^2}}$$

A starting point in statistical tests is to establish a **null hypothesis** (given the symbol $H_0$). This is generally stated, for the *t*-test, in one of the following ways:

- *'The difference between the sample means for* [insert the two variables] *is simply due to random factors, and is not significant.'*
- *'There is no significant difference between the sample means for* [insert the two variables].'

Having carried out a *t*-test, you have your calculated *t*-value. You also know the degrees of freedom for comparing the two samples, $(n_1 + n_2 - 2)$. You are now in a position to use a *t*-table to determine the probability of the null hypothesis being true.

Look across the row for the relevant degrees of freedom (go to the next lower value if the exact value is not included in the table) and find where your calculated *t*-value fits between the *t*-values in the table. Now look up and read off the two *p* values at the top. These are the *p* values within which the probability of the null hypothesis being true lies. A value of $p < 0.05$ means that there is a probability of less than 5 times in

**Exam tip**

You must not state that there is 'no difference'. Of course there is a difference — sample means vary. What you must state is that any difference is not significant or is due to chance.

a hundred of the null hypothesis being true, and so $H_0$ is rejected and a significant difference is concluded.

Obviously there is a chance of being wrong (up to 5 times in a hundred). However, different significance levels are recognised, and you should always quote the level accurately. Table 2 shows the different significance levels.

**Table 2** Different significance levels in statistical tests

| Probability of $H_0$ being accepted ($p$ value) | $p$ value (asterisk indicates significance level) and outcome of test | Significance level |
|---|---|---|
| *greater than* 0.05 | $p > 0.05$, accept $H_0$ | No evidence of significant difference |
| *between* 0.05 *and* 0.01 | $p < 0.05$*, reject $H_0$ | Significantly different, 95% level |
| *between* 0.01 *and* 0.001 | $p < 0.01$**, reject $H_0$ | Highly significantly different, 99% level |
| *less than* 0.001 | $p < 0.001$***, reject $H_0$ | Very highly significantly different, 99.9% level |

You must remember that a $t$-test is not an end in itself. It is simply a tool that allows you to make a decision about significant difference. If you do find a significant difference then you must use your biological understanding to suggest an explanation.

### Worked example

It was considered that limpets (*Patella vulgaris*) when exposed to heavy wave action became squatter, that is shorter in height but with a broader base. This attribute could be measured by calculating the width:height ratio. The table below shows the statistical parameters calculated to compare the width:height ratio for limpets collected at random from an exposed shore and from one sheltered from wave action.

| | Exposed shore | Sheltered shore |
|---|---|---|
| Sample size | 10 | 10 |
| Mean width:height ratio | 2.04 | 1.51 |
| Standard deviation of the mean | 0.114 | 0.032 |

1 The null hypothesis is that there is no significant difference between the width:height ratios of the limpets on the sheltered shore and the exposed shore.

2 Using the formula for $t$:

$$t = \frac{2.04 - 1.51}{\sqrt{0.114^2 + 0.032^2}} = \frac{0.53}{0.118} = 4.492$$

3 The degrees of freedom = 10 + 10 − 2 = 18.

4 Entering $t = 4.492$, at 18 df, in a $t$-table, the probability level for the null hypothesis is determined as $p < 0.001$***.

5 The null hypothesis is rejected and it is concluded that the width:height ratios for limpets from sheltered and exposed shores are (very highly) significantly different.

➡

6 Looking back at the table of data, you can see that the width:height ratio for limpets on the exposed shore is greater — the limpets exposed to wave action tend to be shorter relative to the width of their base. If asked, you might suggest that they have a greater area of 'foot' to hold them against a rock than the limpets on the sheltered shore, meaning that they can grip on to the rock better and are less likely to be washed off by waves; they would also be more streamlined when waves hit them, again making it easier for them to stay attached to the rocks.

## Chi-squared ($\chi^2$) test

In some investigations the DV is a count or frequency (that is, a number of items). This could be the number of fruit fly phenotypes in a genetic cross, or the number of earthworms in a series of fields. The numbers counted (referred to as the **observed frequencies**) will differ at random from those expected on the basis of a reasoned hypothesis. That hypothesis will depend on the investigation: in a genetic cross it may be a 3:1 ratio; in the earthworm counts in, say three fields, equal counts in the fields might be expected. This allows you to calculate the **expected frequencies**.

The **chi-squared ($\chi^2$) test** allows you to decide whether or not the observed frequencies deviate significantly from those expected. The formula for $\chi^2$ is:

$$\chi^2 = \frac{(O - E)^2}{E}$$

You must remember that *the sum of the observed frequencies must equal the sum of the expected frequencies.* Again, a null hypothesis ($H_0$) is set up. A generalised one would be: *'Any differences between the observed frequencies and the expected frequencies are due to chance alone and are not significant.'* However, $H_0$ should be stated specifically for the investigation that you are carrying out, for example:

- For a genetic cross: 'The numbers of wild type and vestigial winged flies only differ from those expected on the basis of a 3:1 ratio as a result of random factors.'
- For the numbers of earthworms in three fields: 'The numbers of earthworms in the three fields are equal and any deviation of the observed counts from this is not significant.'

You use a $\chi^2$ table to determine the probability of this $H_0$ being true. Look across the row for the relevant degrees of freedom, $n - 1$ (where $n$ is the number of categories), and find where your calculated $\chi^2$ value fits between neighbouring tabular $\chi^2$ values; then look up and record the two corresponding $p$ values. These are the $p$ values within which the probability of $H_0$ being true lies.

Note that the chi-squared test can only be used with raw data (the counts); it cannot be used with processed data such as means or percentages.

> **Exam tip**
>
> When quoting $p$ values always ensure that the chevrons are the right way round. Check by ensuring that the larger number is shown as being greater than the lesser number, e.g. $0.05 > p > 0.01$.

### Worked example

Geneticists crossed grey-bodied, straight-winged (wild type) fruit flies (*Drosophila melanogaster*) with flies which were black-bodied and had curled wings. All the $F_1$ flies were wild type, but when these were interbred the $F_2$ flies had a range of phenotypes: 53 were wild type (grey bodies, straight wings), 20 had black bodies, straight wings, 19 had grey bodies, curled wings, and 4 had black bodies, curled wings. Assuming Mendel's second law, the law of independent assortment, the $F_2$ phenotypes would be expected to fulfil a ratio of $9 : 3 : 3 : 1$.

→

1  The null hypothesis is that there is no significant difference between the observed and expected offspring numbers, that is, the results are a good fit to a 9 : 3 : 3 : 1 ratio.

2  The results are entered into a table for the calculation of $\chi^2$:

| Category | O | E | O – E | $(O – E)^2$ | $(O – E)^2/E$ |
|---|---|---|---|---|---|
| Grey body, straight wing | 53 | 54 | –1 | 1 | 0.02 |
| Black body, straight wing | 20 | 18 | 2 | 4 | 0.22 |
| Grey body, curled wing | 19 | 18 | 1 | 1 | 0.06 |
| Black body, curled wing | 4 | 6 | –2 | 4 | 0.67 |

3  Summing the last column provides : $\chi^2 = 0.97$.

4  There are 3 degrees of freedom (4 categories – 1 = 3).

5  The critical value from the $\chi^2$ table (at $p = 0.05$, df = 3) is 7.81.

6  The calculated $\chi^2$ is less than the critical value, so the difference between the expected and observed results is not significant.

7  The results are a good fit to those expected on the basis of the independent inheritance of the characters for body colour and wing shape.

# ◾A2 Unit 3 assessing practical skills

A2 Unit 3 includes a series of internally assessed practical tasks, and a 1 hour 15 minute written examination assessing practical skills. Together these assessments are allocated 75 marks and contribute 12% to the final A-level outcome.

Overall in the A2 Unit 3 assessments, the approximate marks allocated for each assessment objective (AO) are:

AO1   Knowledge and understanding: 27

AO2   Application of knowledge and understanding: 27

AO3   Analysis, interpretation and evaluation of scientific information, ideas and evidence: 21

## Practical tasks internally assessed

You will be asked to complete at least five practical tasks selected by your teacher (from a list provided on p. 78 of the specification). These tasks will allow your implementation skills to be assessed.

You must make a record of the practical tasks that you have carried out. Your work will be assessed by your teacher and made available for presentation to CCEA for moderation. Each task will earn up to 3 marks so that in total 15 marks are available. This contributes 2.4% to the final A-level outcome.

# Practical skills written examination

The A2 Unit 3 examination contributes 9.6% to the final A-level outcome. The paper lasts 1 hour 15 minutes and is worth 60 marks. It consists of between 8 and 10 structured questions, which will vary in length and style. Questions will assess your understanding of A2 practical skills and your ability to apply them to familiar and unfamiliar contexts. A question may also assess the skills required to write a short bibliography detailing source materials.

A bibliography is a list of references — sourced from a book, an article in a scientific journal, a newspaper report, a website — used in an essay or presented in a scientific report. There are a number of formats for listing references and it does not matter which system you use so long as you do so consistently. The standard format for laying them out in a biological science is as follows:

- author name — surname first then initials, followed by other authors, surname first then initials
- date of publication — in parentheses
- title of publication or scientific article
- publishing company, or scientific journal (with volume and page numbers) or website

For example, for this book:

Campton, J. (2018), *CCEA AS/A2 Biology Student Guide: Unit 3 Practical Skills in Biology*, Hodder Education.

For a scientific journal article:

Falconer, A. C. and Hayes, L. J. (1986), 'The extraction and partial purification of bacterial DNA as a practical exercise for GCE Advanced level students', *Journal of Biological Education*, 20 (1) 25–26.

For a website:

Nuffield Foundation, *Practical Biology* (viewed 23/5/2017), www.nuffieldfoundation.org/practical-biology.

The bibliography should have the authors listed alphabetically.

# ■Questions & Answers

This section consists of questions covering a range of different practical tasks. Questions such as these may appear in the exam papers for Unit 1 or Unit 2, though the Unit 3 written paper is designated for assessing 'Practical Skills'.

Following each question, there are answers provided by two students of differing ability. Student A consistently performs at grade A/B standard, allowing you to see what high-grade answers look like. Student B makes a lot of mistakes — ones that examiners often encounter — and grades vary between C/D and E/U.

Each question is followed by a brief analysis of what to look out for when answering the question (shown by the icon ⓔ). All student responses are then followed by comments (preceded by the icon ⓔ). They provide the correct answers and indicate where difficulties for the student occurred, including lack of detail, lack of clarity, misconceptions, irrelevance, poor reading of questions and mistaken meanings of examination terms.

## Microbiology

### Question 1 Aseptic technique and antibiotics

**(a)** Describe how you would use aseptic techniques to produce a spread plate given: a McCartney bottle containing a broth culture of the bacterium *Bacillus subtilis*; a sterile Petri dish containing nutrient agar; sterile Pasteur pipette and spreader; bench-space already swabbed to sterilise its surface; Bunsen burner. (5 marks)

**(b)** Suggest why *Bacillus subtilis* was selected for use in this experiment. (1 mark)

Two spread plates of *Bacillus subtilis* were prepared and an e-strip, containing a different antibiotic, added to each. An e-strip contains different concentrations (as μg cm⁻³) of the antibiotic. The plates were incubated at 25°C for three days. The results are shown in the diagram below.

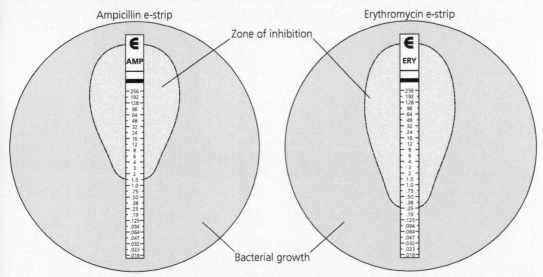

**(c)** Why were the plates not incubated at a temperature higher than 25°C? (1 mark)

**(d)** The results show that for each antibiotic used there were zones of inhibition where bacteria failed to grow. What can you deduce about the effectiveness of the two antibiotics? (3 marks)

**(e)** Why should you be cautious about concluding that one antibiotic might be more effective than another? (1 mark)

Total: 11 marks

*e* This is a fairly straightforward question on testing the effectiveness of antibiotics. Read each question carefully and ensure that answers include sufficient points to meet the marks allocated.

---

### Student A

**(a)** The stopper of the McCartney bottle is removed with the little finger curled towards the palm, and the neck flamed using a Bunsen flame. ✓ A sample is removed from the broth culture of bacteria using a sterile Pasteur pipette. ✓ A few drops are placed onto the surface of the agar, lifting the Petri dish lid only on one side for a minimum amount of time. ✓ Transfer of microorganisms must be done close to the Bunsen burner where air currents draw any airborne microbes upwards. ✓ A sterile spreader is then used to spread the inoculum over the entire surface. ✓ a

**(b)** Because it is not regarded as a bacterium with any pathogenic properties. ✓ a

**(c)** To avoid culturing pathogenic bacteria which thrive in higher temperatures. ✓ a

**(d)** Erythromycin is more effective than ampicillin with the bacterium used. ✓ The minimum inhibitory concentration, MIC, for ampicillin is 1.5, ✓ while the MIC for erythromycin is 0.25. ✓ a

**(e)** Because the same results might not be achieved if a different bacterium was used. ✓ a

*e* **11/11 marks awarded** a Full and correct answers throughout.

---

### Student B

**(a)** The neck of the bacterial culture bottle is sterilised in the Bunsen flame having removed the bung with the little finger. ✓ A wire loop is sterilised in a Bunsen flame and used to remove a sample from the bacterial culture. ✗ After removing a sample, the lid of the culture bottle is flamed again and the bung replaced. ✓ The lid of the Petri dish is lifted only on one side to allow the agar to be inoculated. ✓ The inoculum is smeared backwards and forwards across the surface of one side of the agar and then repeated three times having turned the Petri dish 90° on each occasion. ✗ a

**(b)** Because it is approved for use in schools by the Society for General Microbiology. ✓ b

**(c)** This is the optimum temperature for bacterial growth. ✗ c

**(d)** Erythromycin creates a bigger inhibitory zone and so is the more effective antibiotic. ✓ d

**(e)** Because the effectiveness of an antibiotic would depend on the strain of bacterium. ✓ e

e **6/11 marks awarded** a Student B has described 'streak plating' yet still gains 3 marks for valid points about the use of aseptic techniques. b Correct. c Incorrect: many bacteria, especially pathogens, grow optimally at 37°C, human body temperature. d Correct for 1 mark, but more regarding the data, particularly the MICs of each antibiotic, would be expected. e Correct.

## Question 2 Antimicrobials in plant extracts: 95% confidence limits

The filamentous fungus *Fusarium oxysporum* is a pathogen of many plants, though some plant species possess natural fungicides that help to protect them against infection. The aqueous extracts of four plants were investigated for their antifungal properties: parsley (*Petroselinum crispum*), rosemary (*Rosmarinus officinalis*), onion (*Allium cepa*) and garlic (*Allium sativum*). Antifungal activity was determined using the agar well diffusion method. Petri dishes containing potato dextrose agar were prepared, inoculated with a block of *Fusarium oxysporum* and a standard volume of plant extract added to a well on the opposite side, as shown below.

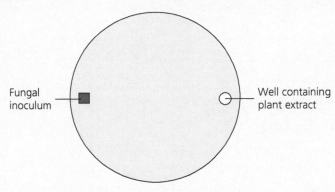

For each plant species tested, 10 replicates were prepared and the agar plates incubated at 25°C for 3 days. Lack of fungi in the vicinity of the well indicates antifungal activity and the area of this inhibition zone was measured for each plate. From the data for each plant extract, the mean and standard deviation (error) of the mean were calculated and for three of the plant species, the 95% confidence limits were determined. The means and 95% confidence limits are shown in the bar graph overleaf.

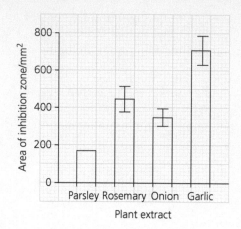

(a) For parsley, the mean ($\bar{x}$) was 170 mm$^2$, while the standard deviation of the mean ($\hat{\sigma}_{\bar{x}}$) was 32. Calculate the 95% confidence limits for parsley. Show your working. (3 marks)

(b) Identify which of the means for the four plant species is the most reliable. Explain your reasoning. (2 marks)

(c) Compare the mean and confidence limits for rosemary, onion and garlic. Do these suggest that there is a significant difference in the effectiveness of the plant extracts in providing antifungal activity? Explain your reasoning. (3 marks)

Total: 8 marks

ⓔ While presented in the context of plant extracts' antimicrobial properties, this is a standard question testing your understanding of sample means and their 95% confidence limits.

---

**Student A**

(a) At 9 degrees of freedom and $p = 0.05$, $t = 2.262$. ✓

95% confidence limits are provided by $170 \pm 2.262 \times 32 = 170 \pm 72.38$ ✓

Lower limit = 97.6, while upper limit = 242.4 ✓ ⓐ

(b) The mean for the onion extract is the most reliable as it has the narrowest limits. ✓ This suggests that its sample mean is closer to its true mean, than for the other plant extracts. ✓ ⓑ

(c) The 95% confidence limits for the means of both rosemary and onion do not overlap with the 95% confidence limits for garlic, ✓ suggesting that the garlic extract has a significantly higher antifungal effect. ✓ ⓒ

---

ⓔ **7/8 marks awarded** ⓐ A correct and well-presented calculation. ⓑ A full answer. ⓒ 2 marks are awarded for a thorough comparison of rosemary and onion with garlic extract, but Student A has failed to compare rosemary with onion.

**Student B**

(a) 95% CL = 170 ± 2 × 32 ✗

Upper limit is 234 and lower limit is 106 ✓ a

(b) The 95% CL for garlic extract suggest that it is significantly different from the other plant extracts. ✗ c

(c) The confidence limits for garlic do not overlap with those for either onion or rosemary suggesting significant difference. ✓ The 95% confidence limits for rosemary and onion overlap indicating that there is no significant difference. ✓ e

ⓔ **3/8 marks awarded** a Student B has failed to show how the appropriate *t* value is determined and no mark is awarded for using an incorrect value. However, there is no further penalty in providing upper and lower limits based on the *t* value used. b The student has confused comparing confidence limits to suggest significant difference with using the width of the confidence limits in determining the reliability of a mean. c This gains 2 marks but for a third mark reference should have been made to the mean for garlic extract being significantly *greater* and so having more effective antifungal properties.

# Coordination and control

## Question 3 Gibberellins and plant growth: a *t*-test

An experiment was carried out on the effect of gibberellic acid (a gibberellin) using young pea seedlings. Pea (*Pisum sativum*, variety Feltham First) seeds were soaked for 8 hours, sown in compost and grown in a greenhouse. An experimental group of 40 seeds were soaked in 0.01% gibberellic acid, while a control group were soaked in water. Fifteen days after sowing, 20 seedlings were randomly selected from each group and their third internode measured (the length between the seed and the first node was the first internode). From the measurements, the mean and standard deviation of the mean for each group were calculated and presented in the table below.

| Statistical parameters relating to internode length of pea seedlings/mm | Treatment | |
| --- | --- | --- |
| | Seeds were soaked in 0.01% gibberellic acid | Seeds were soaked in water |
| Mean ($\bar{x}$) | 60.2 | 42.3 |
| Standard deviation of the mean ($\hat{\sigma}_{\bar{x}}$) | 6.4 | 4.5 |

(a) Describe how you would measure the third internode. (1 mark)

(b) Identify two factors which were controlled in the experiment. (2 marks)

(c) A *t*-test was undertaken to compare the mean growth for the two groups of pea plants.

    (i) Calculate the value of *t* for a comparison of the two means. (3 marks)

(ii) State the following:
- the null hypothesis for this test
- the degrees of freedom
- the probability value for the calculated $t$
- your decision about the null hypothesis

(4 marks)

(d) A student wished to investigate the effect of a mixture of an auxin (IAA) and a gibberellin (gibberellic acid) together on pea seedling growth. What controls should the student include in the experimental design?

(2 marks)

Total: 12 marks

ⓔ It is unlikely that you will have investigated the effect of gibberellic acid, though this question is about undertaking a comparison of sample means using a $t$-test, a statistical skill which you will have practised.

---

### Student A

(a) This is the distance between the second and third node. ✓ ⓐ

(b) The same strain of pea, Feltham First, was used. ✓ The two sets of pea seeds were soaked beforehand for the same length of time. ✓ ⓐ

(c) (i) $t = \dfrac{60.2 - 42.3}{\sqrt{6.4^2 + 4.5^2}}$ ✓ $= \dfrac{17.9}{\sqrt{40.96 + 20.25}}$ ✓ $= \dfrac{17.9}{7.824} = 2.288$ ✓ ⓐ

(ii) Null hypothesis ($H_o$) is that the difference between the sample means is due entirely to chance, i.e. is not significant. ✓ Degrees of freedom = 38. ✓ The probability level for the $t$ value is $0.05 > p > 0.01$. ✓ The difference between the means is significantly different. ✓ ⓐ

(d) Apart from a control in which seeds were only immersed in water, the investigation requires a control in which seeds are treated with gibberellic acid only ✓ and a control in which seeds are treated with IAA only. ✓ ⓐ

ⓔ **12/12 marks awarded** ⓐ Well-phrased and accurate answers for full marks.

---

### Student B

(a) This is the distance between two nodes. ✗ ⓐ

(b) The seeds were grown in compost for the same length of time, 15 days. ✓ The temperature was kept constant. ✗ ⓑ

(c) (i) $t = \dfrac{60.2 - 42.3}{\sqrt{6.4^2 + 4.5^2}}$ ✓ $= \dfrac{17.9}{\sqrt{61.21}}$ ✓ $= 2.288$ ✓ ⓒ

(ii) The null hypothesis is there is no difference between the means. ✗ The degrees of freedom for the test are 78. ✗ The probability level for the $t$ value is $0.05 > p > 0.01$. ✓ The difference between the means is significant. ✓ ⓓ

(d) Controls are required where pea seeds are treated with gibberellin only ✓ and with auxin only. ✓ ⓔ

**ⓔ 8/12 marks awarded** **ⓐ** Student B has identified an internode but has failed to use the information in the question — that the first internode is the distance between the seed and the first node. **ⓑ** The first statement is correct but, while the seedlings were grown in a greenhouse, there is no evidence that temperature was kept constant. **ⓒ** The calculation of *t* is correct. **ⓓ** The null hypothesis is not that there is no difference but that it is not significant. The student has not noticed that only 20 of the 40 seeds in each group were measured so that the wrong degrees of freedom are quoted. The *p* value and the decision of significant difference are correct. **ⓔ** Good understanding of controls for 2 marks.

## Question 4 The mammalian eye

**The photomicrograph below shows a section through a mammalian eye. Note that this section does not pass through the pupil.**

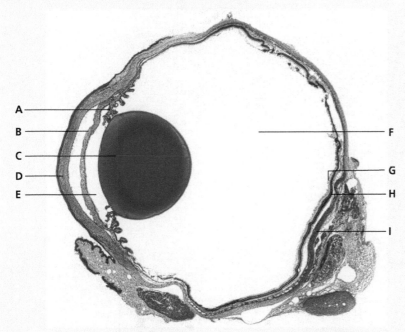

**Identify the features labelled A to I.**                    (9 marks)

Total: 9 marks

**ⓔ** This question requires you to demonstrate a comprehensive understanding of eye structure. Study the photomicrograph thoroughly before answering.

---

**Student A**

A ciliary body; ✓ B iris; ✓ C lens; ✓ D conjunctiva; ✗ E aqueous humour; ✓
F vitreous humour; ✓ G retina; ✓ H choroid; ✓ I sclera. ✓ **ⓐ**

---

**ⓔ 8/9 marks awarded** **ⓐ** These are correct except for D which is the cornea — the conjunctiva is on the outer surface.

e **5/9 marks awarded** a Five features are correct. However, for A the spelling is too poor to be allowed a mark; B cannot be the pupil, which is simply a space, and you are informed that the section does not go through the pupil; and E and F have been identified the wrong way round.

## Question 5 Muscle types

**The drawings below show three types of muscle.**

A 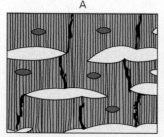 B 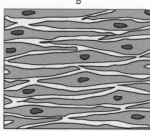 C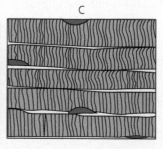

(a)  Identify each of the muscle types, A, B and C.                    (3 marks)

(b)  Which of the muscle types would you find in:
  ● **muscularis mucosa**
  ● **sinoatrial node**                                                (2 marks)

Total: 5 marks

e This is a straightforward question as long as you know the different muscle types.

**Student A**

(a)  A cardiac muscle; ✓ B smooth muscle; ✓ C skeletal muscle. ✓ a

(b)  The muscularis mucosa contains smooth muscle. ✓ The sinoatrial node contains cardiac muscle. ✓ a

e **5/5 marks awarded** a Full marks.

**Student B**

(a)  A cardiac muscle; ✓ B smooth muscle; ✓ C striated muscle. ✗ a

(b)  The muscularis mucosa contains smooth muscle. ✓ The sinoatrial node contains striated muscle. ✗ a

e **3/5 marks awarded** a Both skeletal and cardiac muscle are striated so this cannot be allowed as an answer.

# Populations

## Question 6 Calculating cell density using a haemocytometer

An experiment was set up to investigate the growth of a yeast population. A small number of yeast cells were allowed to multiply in a nutrient medium. The number of cells was estimated by removing a sample of the population at regular intervals and counting the yeast cells using a haemocytometer.

(a)  The samples removed need to be representative of the yeast population at the time of sampling. Suggest how this is achieved. (1 mark)

A sample from the culture, at an early stage of population growth, was taken and added to the haemocytometer slide. The diagram below shows the yeast cells in one type-B square. The distance between the surface of the type-B square and the overlying coverslip is 0.1 mm.

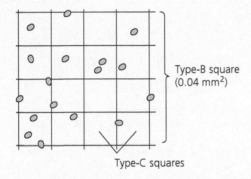

Type-B square
(0.04 mm$^2$)

Type-C squares

(b)  Suggest why type-B squares, rather than type-C squares, were used to estimate the number of yeast cells in the sample. (1 mark)

(c)  When counting cells within a square, what procedure is used to account for those cells touching the boundaries? Explain why this procedure is valid to ensure accurate results. (2 marks)

(d)  On the basis of your answer to (c), how many yeast cells are present in the type-B square? (1 mark)

(e)  Using your answer in (d), calculate the number of yeast cells per mm$^3$. Show your working. (2 marks)

(f)  State one precaution needed, when adding a sample to the haemocytometer slide, to ensure accuracy of the result. (1 mark)

(g)  In older populations of yeast there may be too many yeast cells to clearly see the grid lines on the haemocytometer slide. Suggest how such a dense population might be treated to enable a count to be made, and how this count would subsequently be used to obtain an estimate of the population size. (2 marks)

Total: 10 marks

# Questions & Answers

*e* This is a question about a practical that you will have undertaken. Apart from knowing how to use the haemocytometer to make a cell count, you must understand important precautions required during the procedure.

---

**Student A**

**(a)** The culture of yeast is shaken so mixing the cells throughout. ✓ a

**(b)** Some type-C squares are empty. ✓ a

**(c)** Count cells touching the top and left sides of a type-B square and ignore those touching the bottom and right sides. ✓ Otherwise cells might be counted twice. ✓ a

**(d)** 15 ✓ a

**(e)** 15 cells in $0.04\,mm^2 \times 0.1\,mm$ ✓ = 15 cells in $0.004\,mm^3$ = 3750 cells $mm^{-3}$. ✓ a

**(f)** Make sure that none of the sample enters the grooves on the haemocytometer, ✓ otherwise the coverslip will float over the grid area. a

**(g)** A sample from the yeast culture would be diluted, say × 10, ✓ and the final population estimate multiplied by the dilution factor. ✓ a

---

*e* **10/10 marks awarded** a Full marks from a student who has obviously undertaken the practical.

---

**Student B**

**(a)** The yeast population is swirled before a sample is removed. ✓ a

**(b)** To ensure accuracy of the results. ✗ b

**(c)** Use the north-west rule. ✓ c

**(d)** 15 ✓ d

**(e)** 15 cells in $0.04\,mm^2 \times 0.1\,mm$ ✓ = 375 cells $mm^{-3}$ ✗ e

**(f)** When the sample is added alongside the coverslip it must be allowed to enter the space by capillarity. ✓ f

**(g)** Dilute the sample so reducing the density of the yeast cells on the grid. ✓ g

---

*e* **6/10 marks awarded** a Correct. b This is too vague. An acceptable answer would have been 'there are small numbers of cells overall'. c This is acceptable for including cells at the top and left boundaries. However, Student B has not explained the need for this rule. d Correct. e The initial understanding is correct but the final calculation is wrong. f Correct. g Understanding the need for dilution is awarded 1 mark, but the student has forgotten to add that the count would need to be multiplied by the dilution factor.

# Question 7 Estimating population size using the capture–mark–recapture technique

A population of garden snails (*Helix aspersa*) inhabit a stone wall. To estimate the population size of the snails, a capture–mark–recapture technique was used.

A sample of 88 snails were captured and marked with Tipp-Ex on the underside of their shells.

**(a)** Suggest why the snails were marked on the underside of their shells.                    (1 mark)

Three days after the release of the marked snails, a second sample was captured. Of these, 33 were found to be marked and 48 were not.

**(b)** Explain why there was a 3-day interval between the release of the marked snails and further sampling of the population.                    (1 mark)

**(c)** Calculate the estimated size of the snail population. Show your working.                    (3 marks)

**(d)** Explain what would happen to the estimate of population size if the marked snails, upon release, simply 'hid' in the crevices of the stone wall and so were unlikely to have been caught after 3 days.                    (2 marks)

Total: 7 marks

ⓔ This question is about another practical that you should have undertaken. Part (d) is trickier but think carefully and you can work out the answer.

---

**Student A**

**(a)** So that the marked snails would not be more obvious to a potential predator. ✓ a

**(b)** Enables marked snails to redistribute themselves within the population. ✓ a

**(c)** The estimate of population size $= \dfrac{s_1 \times s_2}{r}$ ✓ $= \dfrac{88 \times 81}{33}$ ✓ $= 216$ ✓ a

**(d)** The number of recaptures would be underestimated, ✓ and so the population size would be overestimated. ✓ a

---

ⓔ **7/7 marks awarded** a Full marks.

**Student B**

**(a)** So that marked snails could be recognised in a subsequent sample. ✗ a

**(b)** So that the subsequent sample is representative of the population. ✓ b

**(c)** $N = \dfrac{s_1 \times s_2}{r}$ ✓ $= \dfrac{88 \times 48}{33}$ ✗ $= 128$ ✓ c

**(d)** The estimate of population size would be unreliable. ✓ d

---

ⓔ **4/7 marks awarded** a Student B has failed to explain why the snails were marked on the 'underside'. b Correct. c The student gets a mark for the correct formula, but applies the wrong values — the second sample consists of the marked and unmarked snails — though the final arithmetic is appropriate for the values given. d This is fine for 1 mark — see Student A for a full answer.

# Respiration and photosynthesis

## Question 8 Calculating RQ values using a respirometer

A student investigated the respiration of wheat seeds using the apparatus shown in the diagram below.

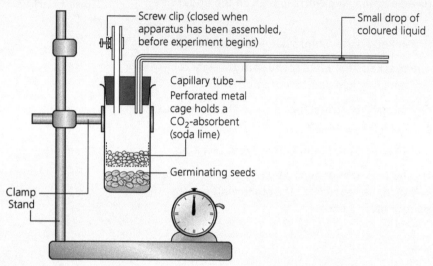

(a) In the set-up shown above, the drop of coloured liquid moved to the left by 40 mm in 10 minutes. The volume of a cylinder = $\pi r^2 l$ where $\pi = 3.14$, $r$ is the radius of the capillary tube bore and $l$ is the distance moved by the coloured liquid. The capillary tube bore had a diameter of 1.1 mm. The mass of germinating seeds used was 8 grams.

(i) Calculate the oxygen consumption of the germinating seeds as $mm^3 g^{-1} min^{-1}$. Show your working. (3 marks)

(ii) Explain how you would use the apparatus to measure carbon dioxide production of the germinating seeds. (3 marks)

(b) The student was given the results of a previous investigation — see table below.

| | Volume of $O_2$ consumed /$mm^3 hour^{-1}$ | Volume of $CO_2$ produced /$mm^3 hour^{-1}$ |
|---|---|---|
| Seeds that had just started to germinate | 220 | 156 |
| Seeds that had been germinating for 24 hours | 225 | 220 |

(i) Calculate the RQ values for the seeds at each of the stages of germination. (2 marks)

(ii) What can you deduce from the RQ values calculated? (3 marks)

(c) State two precautions needed in using a respirometer to ensure accurate results. (2 marks)

Total: 13 marks

ⓔ In this question on respirometry you will need to show understanding of both practical and theoretical aspects. As always, read the information supplied and the questions carefully.

---

**Student A**

**(a) (i)** Rate of oxygen consumption = $40\,mm \times 3.14 \times 0.55^2\,mm^2$ per $8\,$grams in $10\,$minutes ✓ = $38\,mm^3$ per $8\,g$ in $10\,$minutes = $4.75\,mm^3\,g^{-1}$ in $10\,$minutes ✓ = $0.475\,mm^3\,g^{-1}\,min^{-1}$ ✓ ⓐ

**(ii)** The soda lime is removed. ✓ The difference between the position of the drop of liquid with and without soda lime indicates the volume of $CO_2$ produced. ✓ If less $CO_2$ is produced than $O_2$ consumed, then the droplet will move a little to the left. ✓ ⓐ

**(b) (i)** At the start of germination, RQ = 0.71. ✓ After 24 hours, RQ = 0.98. ✓ ⓐ

**(ii)** Respiration is aerobic throughout, ✓ though initially the substrate used is lipid, ✓ while after 24 hours it changes to carbohydrate. ✓ ⓐ

**(c)** The apparatus must be airtight, otherwise gases could enter and escape. ✓ Handling the apparatus must be avoided, otherwise a temperature increase would cause gases in the apparatus to expand. ✓ ⓐ

---

ⓔ **13/13 marks awarded** ⓐ Calculations are correct and well set out. Explanations are complete and well-phrased.

---

**Student B**

**(a) (i)** $O_2$ consumption = $40 \times 3.14 \times 1.1^2\,mm^3$ per $10\,$minutes ✗ = $152\,mm^3$ per $10\,$minutes ✗ = $1.52\,mm^3\,min^{-1}$ ✓ ⓐ

**(ii)** The same apparatus is used without the soda lime. ✓ Movement of the drop of liquid is due to the difference in CO2 production and O2 consumption. ✓ ⓑ

**(b) (i)** The RQ is 0.709 at the start, ✓ and 0.978 after 24 hours. ✓ ⓒ

**(ii)** At the start the substrate respired is lipid, ✓ and after 24 hours the substrate used is carbohydrate. ✓ ⓓ

**(c)** If the bung wasn't put on properly some oxygen could get into the respirometer. ✓ Ensure that movement of the drop of liquid in the capillary tube is accurately measured. ✗ ⓔ

---

ⓔ **8/13 marks awarded** ⓐ Student B has made two mistakes: the diameter for the capillary bore is used rather than the radius; while the mass of seeds, 8 g, has been ignored. However, these mistakes are not carried forward and 1 mark is awarded. ⓑ 2 marks are awarded, though the student has not fully explained how carbon dioxide production is calculated. ⓒ Both RQ values are correct. ⓓ The link between RQ and substrate is well understood, but there is no mention that respiration is aerobic. ⓔ One correct precaution identified.

# Question 9 Chromatography of plant pigments

Algae are photosynthetic organisms that contain a number of plant pigments. An extract was obtained from the cells of a species of alga and the pigments separated using two-way chromatography. In two-way chromatography, the extract is spotted in one corner of the sheet, run in the first solvent along one side of the sheet, after which the sheet is rotated 90° and run with a second solvent in the new direction, so that the resultant spots are spread over the entire sheet. The chromatogram produced is shown below.

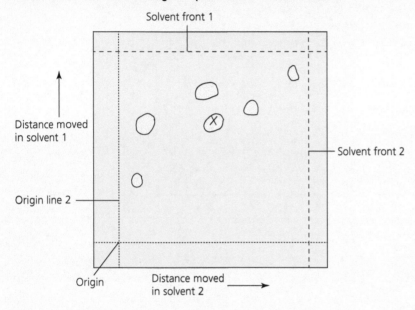

The table below shows the $R_f$ values for the pigments on the chromatogram in each of the solvents used.

| | Pigment | | | | | |
|---|---|---|---|---|---|---|
| | Carotene | Chlorophyll a | Chlorophyll b | Lutein | Neoxanthin | Violaxanthin |
| $R_f$ in solvent 1 | 0.96 | 0.84 | 0.65 | 0.74 | 0.32 | 0.65 |
| $R_f$ in solvent 2 | 0.96 | 0.47 | 0.10 | 0.73 | 0.05 | 0.48 |

(a) Using information in the table, explain why two-way chromatography was necessary to separate all of the pigments in the extract. (3 marks)

(b) Name the pigment labelled X on the chromatogram. Show how you arrived at your answer. (3 marks)

Total: 6 marks

ⓔ You may not have undertaken two-way chromatography, but there is sufficient information supplied for you to understand the procedure. This is testing your ability to apply your understanding.

**Student A**

**(a)** Because while the other pigments are separated in solvent 1, ✓ chlorophyll *b* and violaxanthin have the same $R_f$ value. ✓ Only in solvent 2 are chlorophyll *b* and violaxanthin separated. ✓ ⓐ

**(b)** X is violaxanthin. ✓ Its $R_f$ value in solvent 1 is approximately $\dfrac{32}{50} = 0.64$, ✓ while in solvent 2 its $R_f$ value is approximately $\dfrac{25}{50} = 0.50$. ✓ ⓐ

ⓔ **6/6 marks awarded** ⓐ Answers have been well worked out — full marks.

**Student B**

**(a)** Chlorophyll *b* and violaxanthin ✓ have the same $R_f$ value in the first solvent. ✓ The second solvent on its own would have the same problem ✓ with chlorophyll *a* and violaxanthin. ⓐ

**(b)** The $R_f$ value in solvent 1 is 0.5, ✗ while in solvent 2 it is 0.64. ✗ However, there is no pigment corresponding to these values. ✗ ⓑ

ⓔ **3/6 marks awarded** ⓐ Well explained for 3 marks. ⓑ Unfortunately, Student B has mixed up the two solvent fronts and has failed to score. Since there did not appear to be a pigment with these $R_f$ values, Student B should have checked the calculations again.

## Question 10 Identifying hydrogen acceptors using a redox indicator

**(a)** In an investigation of yeast population growth, a student wished to determine if the cells were living or dead. To achieve this, the student added some drops of dilute methylene blue to the samples that were being examined with a haemocytometer. Dead cells appeared blue while living cells were white.

Explain why methylene blue distinguishes dead from living cells. (2 marks)

**(b)** In an investigation of the role of chloroplasts in the light-dependent reaction of photosynthesis, a chloroplast suspension was made by grinding fresh leaves in buffer solution and centrifuging the mixture. The chloroplasts were tested for their ability to reduce DCPIP (2,6-dichlorophenol-indophenol), a blue dye, which becomes colourless when reduced. Test tubes were prepared and treated in different ways. The colour of the tube contents was recorded at the start and after 20 minutes. This information is summarised in the table below.

| Test tube | Contents | Treatment | Colour | |
|---|---|---|---|---|
| | | | At start | After 20 minutes |
| A | Chloroplast suspension plus DCPIP | Tube kept in bright light | Blue/green | Green |
| B | Chloroplast suspension plus DCPIP | Tube kept in dark | Blue/green | Blue/green |
| C | Buffer solution plus DCPIP | Tube kept in bright light | Blue | Blue |

(i) Explain the result for test tube A. (3 marks)

(ii) Explain the role of test tube C in the investigation. (1 mark)

(iii) The chloroplast suspension produced by centrifugation may also contain mitochondria. Explain the evidence from test tube B that mitochondria are not responsible for reducing the DCPIP. (2 marks)

(iv) Suggest why conclusions made only on the basis of the data in the table may not be reliable. (1 mark)

Total: 9 marks

ⓔ This question requires an understanding of redox indicators in investigating the activity of chloroplasts and mitochondria though it moves on to become a general experimental design question.

### Student A

(a) Live cells contain mitochondria with dehydrogenase activity releasing hydrogen which reduces the methylene blue so it becomes colourless. ✓ Dead cells have no dehydrogenase activity and no reducing power. ✓ ⓐ

(b) (i) Light-activated chlorophyll within the chloroplasts release electrons, ✓ reducing DCPIP which becomes colourless. ✓

(ii) To show that chloroplasts are responsible for the change. ✓

(iii) Mitochondria will release hydrogen, ✓ which would reduce DCPIP so that tube B would also become green. ✓

(iv) There is only one set of results so the whole experiment should be repeated. ✓ ⓐ

ⓔ **8/9 marks awarded** ⓐ This would have gained full marks except that the answer to (b) (i) is incomplete. Student A should have explained the resultant green colour by stating that after DCPIP is reduced 'only the colour of the chloroplasts can be seen'.

### Student B

(a) The cells will appear differently when viewed under the microscope — living cells are white, while dead cells are blue. ✗ ⓐ

(b) (i) In chloroplasts chlorophyll molecules are excited by light to release electrons. ✓ DCPIP is reduced, ✓ becoming colourless so that only the green colour of the chloroplasts can be seen. ✓ ⓑ

(ii) This is a control. ✗ ⓒ

(iii) Tube B shows light is necessary for colour change. ✓ Mitochondria do not have a pigment to absorb light. ✓ ⓓ

(iv) There need to be more results. ✗ ⓔ

ⓔ **5/9 marks awarded** ⓐ This is just a repeat of the information in the question, not an explanation for the difference. ⓑ A full answer for 3 marks. ⓒ It is not sufficient to say it is a control. The student needed to say how it acts as a control — see Student A's answer. ⓓ This is correct and an alternative answer to that of Student A. ⓔ This is too vague.

# DNA technology

## Question 11 DNA extraction and separation using gel electrophoresis

**(a)** Outline how a sample of DNA may be extracted from the tissue of an onion.                  (4 marks)

**(b)** A sample of DNA was treated with an enzyme to cut it into fragments. The DNA fragments were then added to a well, prepared in a gel electrophoresis apparatus, and an electrical current applied.

  **(i)** Name the type of enzyme used to cut the DNA into fragments.                  (1 mark)

  **(ii)** The well containing the DNA is located at the negative (cathode) end of the apparatus and the DNA moves in the gel towards the positive (anode) end. Explain why.                  (1 mark)

  **(iii)** As the DNA fragments moved through the gel they became separated. Explain why.                  (2 marks)

  **(iv)** How might the different bands of separated DNA fragments be located?                  (1 mark)

Total: 9 marks

ⓔ This question tests your knowledge and understanding of two DNA practicals. You will benefit from having undertaken them both.

---

**Student A**

**(a)** Onion is finely chopped and homogenised in a solution of salty detergent. ✓ Homogenisation disrupts the cell walls and detergent ruptures the cell membranes, releasing chromatin. ✓ The extract is filtered and protease enzyme added to digest protein and free DNA from its histone backbone. ✓ The DNA is precipitated by gently adding ice-cold ethanol down the side of the beaker. ✓ Precipitated DNA fibres form at the water–ethanol interface and are drawn out by rotating a fine glass rod. ⓐ

**(b) (i)** Restriction endonucleases. ✓

  **(ii)** Because DNA carries negative charges. ✓

  **(iii)** The fragments will be of different lengths dependent on the location of restriction sites on the DNA. ✓ Smaller fragments will travel further. ✓

  **(iv)** A stain such as Azure A or Toluidine blue is added to allow the DNA bands to be seen. ✓ ⓐ

---

ⓔ **9/9 marks awarded** ⓐ Full marks for well-phrased answers.

ⓔ 5/9 marks awarded ⓐ This gains 3 marks, but there is no attempt to explain the different aspects of the procedure, e.g. that detergent ruptures cell membranes. ⓑ Correct. ⓒ The student needed to say what the charge is. ⓓ This is correct for 1 mark. However, Student B has not explained why there are fragments of different lengths. ⓔ For the mark a dye should be named.

# Inheritance

## Question 12 Analysing a genetic cross: a $\chi^2$ test

In maize (*Zea mais*), the kernels ('seeds') within a cob ('ear') may be purple or yellow and smooth or shrunken. A student was presented with an $F_2$ cob which had been grown from parents, one of which was genetically purple, smooth while the other was yellow, shrunken. The student counted the different kernel types within the cob and recorded the results. A $\chi^2$ test was then undertaken to determine if the results fitted an expected 9:3:3:1 ratio. The results and the table for calculating the $\chi^2$ value are presented in the table opposite.

| | Observed number $(O)$ | Expected ratio | Expected number $(E)$ | $(O-E)^2/E$ |
|---|---|---|---|---|
| Purple, smooth | 189 | 9 | | |
| Purple, shrunken | 68 | 3 | | |
| Yellow, smooth | 54 | 3 | | |
| Yellow, shrunken | 17 | 1 | | |
| Total | 328 | | | $\chi^2 =$ |

(a)   Complete the table and show the value for $\chi^2$.                                    (3 marks)

(b)   State the following:
  • the null hypothesis for this test
  • the degrees of freedom
  • the probability value for the calculated $\chi^2$
  • your decision about the null hypothesis                              (4 marks)

(c)   Explain why a 9:3:3:1 ratio was the expected outcome of the cross.        (2 marks)

Total: 9 marks

@ Practising statistical tests, particularly with respect to genetics problems, should make this question straightforward.

---

**Student A**

**(a)**

|  | O | E | $(O-E)^2/E$ |
|---|---|---|---|
| Purple, smooth | 189 | 184.5 | 0.110 |
| Purple, shrunken | 68 | 61.5 | 0.687 |
| Yellow, smooth | 54 | 61.5 | 0.915 |
| Yellow, shrunken | 17 | 20.5 ✓ | 0.598 ✓ |
| Total | 328 | 328 | $\chi^2 = 2.310$ ✓ |

a

**(b)** The null hypothesis is that the observed results do not differ significantly from those expected, ✓ and that any difference is due to chance. Degrees of freedom = 3. ✓ $0.9 > p > 0.5$. ✓ The null hypothesis is accepted. ✓ a

**(c)** Two genes are involved, each with two alleles, one of which is dominant to the other — purple > yellow and smooth > shrunken. ✓ The two genes are independently inherited. ✓ a

---

@ **9/9 marks awarded** a Full marks for correct answers throughout.

---

**Student B**

**(a)**

|  | O | E | $(O-E)^2/E$ |
|---|---|---|---|
| Purple, smooth | 189 | 185 | 0.086 |
| Purple, shrunken | 68 | 62 | 0.581 |
| Yellow, smooth | 54 | 62 | 1.032 |
| Yellow, shrunken | 17 | 21 ✗ | 0.762 ✓ |
| Total | 328 | 330 | $\chi^2 = 2.461$ ✓ |

a

**(b)** The null hypothesis is that any difference between observed and expected results is due to random factors. ✓ D.f. = 3. ✓ $0.5 > p > 0.1$. ✓ Accept null hypothesis. ✓ b

**(c)** This is a dihybrid cross. ✗ c

---

@ 6/9 marks awarded a Student B has rounded up the expected values which is wrong since $\Sigma O$ does not equal $\Sigma E$. Note that expected values are based on probability and so may have decimals. This error is not carried forward. b All correct, with the $p$ value appropriate for the $\chi^2$ value quoted though different from that of Student A because of the error in rounding up expected values. c This does not sufficiently explain the 9:3:3:1 ratio.

# Kingdom Animalia

## Question 13 Insect mouthparts: a bibliography

**Read the following extracts (A, B and C) about insect mouthparts and answer the questions that follow.**

> **A** In the mouthparts of a locust and an aphid equivalent parts are differently constructed and perform different functions. Thus the locust has serrated mandibles for cutting and chewing whereas in the aphid the mandibles form the sides of the needle-like proboscis through which plant juices are sucked.
>
> Jenny Chapman & Michael Roberts, Cambridge University Press, 1997, Biodiversity

> **B** The mouthparts of a locust are shown below. The mandibles are crushing jaws. The labium and labrum are the lower and upper lip. The maxillae move food into the mouth, and the palpi (plural of palpus) assist in tasting.
>
>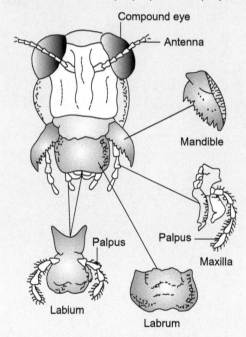
>
> Worth Publishers Inc., 1989, Biology, Helena Curtis & N. Sue Barnes

> **C** Remove the head from a dead locust and boil it gently in a test tube containing a little potassium hydroxide solution until it is soft and semi-transparent. Cut off the labrum (upper lip) which shields the mouthparts. Using forceps, remove mandibles first, then the two maxillae and finally the labium. Examine the mouthparts under low power.
>
> Biology: Students' Manual, M. B. V. Roberts & T. J. King,
> Thomas Nelson Ltd., 1987

**(a) (i)** Aphids insert their proboscis into the phloem of plants to feed on sugars and other nutrients in the sieve tubes. How are the mouthparts of aphids adapted for this function? (1 mark)

**(ii)** How are the mouthparts of a locust adapted for feeding on vegetation? (1 mark)

**(iii)** On which mouthparts of a locust are the sensory organs for tasting to be found? (1 mark)

**(iv)** Suggest why, in preparing a locust for dissection of the mouthparts, the head is heated in dilute potassium hydroxide. (1 mark)

**(v)** Structures that have underlying similarity, suggesting a common evolutionary origin, even when carrying out different functions, are said to be homologous. Describe an example of homologous structures in the mouthparts of insects. (1 mark)

**(b)** Write a bibliography for the three extracts. (3 marks)

Total: 8 marks

ⓔ Read the question carefully and several times if necessary. In part (a), you are not expected to know the answers but use the information provided in the extracts. In part (b), you are expected to use the simple rules for writing a bibliography.

---

**Student A**

**(a) (i)** An aphid has a needle-like proboscis for penetrating plant stems. ✓

**(ii)** A locust has serrated mandibles for crushing vegetation. ✓

**(iii)** Each maxilla and the labium possess palpi. ✓

**(iv)** To soften the tissues to allow the mouthparts to be removed more readily. ✓

**(v)** The mandibles form the basis of the mouthparts of both aphid and locust, that is the same structure is differently adapted for a different function. ✓ ⓐ

**(b)** Chapman, J. & Roberts, M. (1997) Biodiversity, Cambridge University Press. ✓

Curtis, H. & Barnes, N. S. (1989) Biology, Worth Publishers Inc. ✓

Roberts, M. B. V. & King, T. J. (1987) Biology: Students' Manual, Thomas Nelson Ltd. ✓ ⓐ

**e** **8/8 marks awarded** **a** Full marks. Notice that in part (b) the first author has their surname first, then initials; and that the order is author(s), date, title and then publisher.

---

**Student B**

**(a)** **(i)** The aphid has a piercing proboscis. ✓ **a**

**(ii)** The locust has grinding mandibles. ✓ **b**

**(iii)** The palpi. ✗ **c**

**(iv)** This treatment facilitates the dissection of the mouthparts. ✓ **d**

**(v)** Different insects have different mouthparts adapted to their type of feeding. ✗ **e**

**(b)** Chapman, Jenny & Michael Roberts (1997) Biodiversity, Cambridge University Press. ✓

Curtis, Helena & N. Sue Barnes (1989) Biology, Worth Publishers Inc. ✓

Roberts, M. B. V. & T. J. King Biology: Students' Manual, Thomas Nelson Ltd. 1987 ✗ **f**

---

**e** **5/8 marks awarded** **a** Correct. **b** Correct. **c** Palpi are the sensory organs but the question asks for the mouthparts on which they are located. **d** Correct. **e** The student has failed to identify the structure from which the feeding apparatus of different insects is adapted. **f** There is a lack of consistency here. The last reference has the date at the end rather than after the authors, as for the first two references.

## AS Knowledge check answers

**1** Benedict's test

**2** Since the amino acid was spotted 30 mm from the line then it travelled 84 − 30 = 54 mm; the solvent front travelled 150 − 30 = 120 mm from the spotting line; 54 mm divided by 120 mm gives an $R_f$ of 0.45.

**3** Ninhydrin is a carcinogen and so gloves, face mask and goggles must be worn and it should only be sprayed onto the chromatogram in a fume cupboard.

**4** The lipase is left in the water bath for 10 minutes before adding it to the milk (substrate) so that it is at the treatment temperature when the reaction starts.

**5** The independent variable is pH, while rate of reaction (time for reaction to be completed) is the dependent variable.

**6** Control: the volumes and concentrations of the starch and amylase; and the temperature at which the reaction occurs. Timing issues relating to the taking of test samples are discussed in the text.

**7** To ensure that amylase structure has been altered before it reacts with starch.

**8** Heat will cause the volume of oxygen released to increase; and will also speed up the rate of reaction.

**9** Use a water bath (at 37°C).

**10** The higher the concentration of protease, the smaller will be the length of boiled albumen left in the glass tube. At high protease concentration, there are more protease active sites colliding with the albumen so that it is more rapidly digested.

**11** Use different buffer solutions, ranging from pH1 to about pH8; or add different amounts (say drops) of dilute HCl to the solution, using narrow range pH test papers to measure the pH.

**12** Fructose

**13** Production can take place continuously; product is enzyme-free, so purification costs reduced.

**14** 1% (10 ÷ 15 × 1.5%)

**15** Blue solutions most effectively absorb red light provided by a red filter. This allows a greater range of colorimeter readings to be produced.

**16** Benedict's test

**17** Rapid initially, with rate of reaction decreasing over time.

**18** × 20000

**19** 261 μm (98 × 240 μm ÷ 90)

**20** The immersing sucrose solution is open and not under pressure, so $\psi_p = 0$ and $\psi = \psi_s$.

**21** Different cells have different amounts of solutes and those with more solutes have a lower (more negative) solute potential. This influences the cells' water potential and, in consequence, leads to movement of water between cells.

**22** The 'wicks' in the respirometer tubes soak up some KOH and so are used to increase the surface area for the effective absorption of $CO_2$.

**23** Hepatic vein

**24** The line transect records species distribution along the environmental gradient; however, there is no record of the relative abundance of different species.

**25** (a) The percentage cover of the entire lawn is 2%. (b) 2% of the lawn area (9 m × 20 m = 180 m²) is covered, 3.6 m².

**26** The overall coverage equates to six small squares. The percentage coverage of creeping buttercup in the quadrat is 6 × 4% = 24%.

**27** $10^3$ mm³

**28** $2.50 × 10^4$

## A2 Knowledge check answers

**1** Antibiotic A

**2** Ensure that the drying process is effective, since ethanol would kill bacteria; or that the paper itself might influence bacterial growth.

**3** Because cardiac muscle is also striated.

**4** 8.15 × 4000 = 32600 cells mm$^{-3}$.

**5** Yeast sample entering the grooves causes the coverslip to float so that the distance between platform and coverslip becomes greater than 0.1 mm.

**6** The second sample size is 60 (48 + 12) and so population size is estimated as 50 × 60 ÷ 12 = 250.

**7** An RQ of 0.85 is due to aerobic respiration taking place using a mixture of lipid and carbohydrate as respiratory substrate. Protein is only a major respiratory substrate during starvation.

**8** Because the blue colour at the surface is due to oxygen diffusing into solution from the atmosphere.

**9** Different plant pigments absorb slightly different wavelengths of light (which is why they have slightly different colours) so increasing the range of wavelengths absorbed.

**10** Because cells also contain mitochondria which produce hydrogen (from the dehydrogenase activity in respiration) which decolourises DCPIP.

**11** Detergent dissolves the lipids in membranes.

**12** Aids the precipitation of DNA when ethanol is added (by decreasing its solubility in water).

**13** Negative

**14** Mosses

**15** Cnidaria

**16** $\chi^2$ test — analysis of categorical data.

**17** The 95% confidence limits overlap suggesting that there is no significant difference between the mean mass of trout in lakes A and C (though a $t$-test is required to confirm this).

# Index

Note: page numbers in **bold** indicate defined terms.

# Index

populations
    animal population, size
        estimation  82
    questions and answers  103–06
    yeast cultures  80–82
population (in statistics)  **87**
precaution  **7**
presence–absence data  36
probability  **87**
protease activity, enzyme
    concentration  12
protein tests  5

## Q
quadrats, sampling with  31–36

## R
random sampling  **31–32**
rates  38
    calculating  46
ratios  39
redox indicators  83–84, 109–10
reliability  **10**
respiration
    dehydrogenase activity  83–84
    questions and answers  106–07
respirometer
    carbon dioxide production
        28–29
    oxygen consumption  27–28
    questions and answers  67–68,
        106–07
    respiratory quotient (RQ)  83
rounding  38

## S
sample mean, statistical analysis  88
sampling  **31**
    abundance measures  35–36
    devices for  33–35
    procedures  31–33
scaling, graphs  41, 44–45
scatter graphs/scattergrams  41,
    47–48
serial dilution  8–9, 53–54
significant figures  38
Simpson's diversity index  73–74
SI units and derivatives  37
skewed distribution  40
species abundance, estimating
    35–36
stage micrometer  18–19, 58–59
standard deviation  88
standard error of the mean  88
standard form, numbers  39
starch–amylase reaction,
    calorimeter use  13–16
statistics  87–93
streak plating  78–79
Student's $t$-test  90–92, 99–100
surface area calculation  39
sweep nets  35

## T
temperature change/control  7
tissues in plants  29
transect sampling  32–33
transmission electron micrograph
    (TEM)  16–17, 56–57

trends, identifying  45–46
true size calculation  17–18
$t$-test  90–92, 99–100

## U
ultrastructure of cells  16–17
units of measurement  37–38

## V
validity  **5**
vascular tissues, plants  29
veins  30
    of the heart  29–30
volume calculation  39

## W
water baths, temperature control  7
water potential of plant tissue
    21–22

## X
xerophytic leaf  29, 68–70

## Y
yeast growth  80–82, 103–04